Dry Etching Technology for Semiconductors

Translation supervised by Kazuo Nojiri
Translation by Yuki Ikezi

Kazuo Nojiri

Dry Etching Technology for Semiconductors

Springer

Kazuo Nojiri
Lam Research Co., Ltd.
Tokyo, Japan

Based on translation from the Japanese language edition: HAJIMETE NO HANDOTAI DRY ETCHING GIJUTSU by Kazuo Nojiri

ISBN 978-3-319-10294-8 ISBN 978-3-319-10295-5 (eBook)
DOI 10.1007/978-3-319-10295-5
Springer Cham Heidelberg New York Dordrecht London

Library of Congress Control Number: 2014949262

Printed on acid-free paper

Springer is part of Springer Science+Business Media (www.springer.com)

Preface

Dry etching is a key technology comparable in importance to lithography as a means for scaling and enhancing the integration level of semiconductor devices. The number of engineers engaged in dry etching development is also comparable to those working on lithography. Lithography technology is relatively easy to understand because resolution is determined by the optical wavelength and the numerical aperture (NA) of a lens. On the other hand, dry etching technology is difficult to understand because of the complicated phenomena taking place inside the etch chamber. Dry etching also requires comprehensive knowledge in electricity, physics, and chemistry because plasma-based etching is driven by physical–chemical reactions. Engineers engaged in dry etching tend to rely on their experience and intuition with their work. Too often, engineers are thrown in to do their work without first gaining an adequate understanding and knowledge of, for example, how to achieve an anisotropic etching, why Cl_2 and HBr are used for silicon etching, why fluorocarbon gases are used for SiO_2 etching, and why high density plasma, such as inductively coupled plasma (ICP), is used for poly-Si and Al etching, while a narrow-gap parallel-plate etcher with medium density plasma is used for SiO_2 etching. Sometimes, even an expert in dry etching may not understand these issues well enough.

Dry etching technology may sometimes be eclipsed by lithography, but, as already mentioned, it is a key technology that is just as critical as lithography. By way of example, in dry etching, (1) specific equipment and process technologies are used for each material, such as Si, SiO_2, and metal; (2) new technologies are constantly being developed, such as Cu damascene processing and other new material processing; (3) plasma damages caused by charged particles lead directly to device yield losses, and it is necessary to understand their mechanism and solutions; and (4) dry etching is even more critical than lithography with double patterning technology that is a hot topic of the day because it determines dimensional accuracy and uniformity. Engineers who are engaged in such a process technology, which continues to become more diverse and evermore sophisticated in order to support an increased range of materials, should gain an adequate understanding of dry etching technology.

This book follows a unique approach that differs from existing publications, helping readers understand the basics of dry etching and its applications. Many books on dry etching focus on complicated plasma theories, or only offer long lists of data related to dry etching. This book will avoid mathematical equations as much as possible, and has been written to make dry etching mechanisms easy to understand. It is also structured to allow readers to systematically learn about the process itself and then about the equipment and the new technologies. There is a chapter dedicated to plasma damages that presents a holistic picture on the topic.

The book is designed not only to help readers understand the fundamentals of dry etching but also gain a more practical knowledge. The author hopes to provide a guiding principle for engineers who pursue dry etching development.

Tokyo, Japan
October 2014

Kazuo Nojiri

English Translation

Translation supervised by **Kazuo Nojiri**
Translation by **Yuki Ikezi**

Contents

Author Bios

Kazuo Nojiri is a CTO of Lam Research Japan. He has 40 years of experience in semiconductor industry. Prior to joining Lam in 2000, he worked for Hitachi Ltd. for 25 years, where he held numerous management positions in dry etching and device integration. He is also known as a pioneer in the research field of charging damage. He published 38 technical papers and three books. In 1984 he was awarded the Okouchi Memorial Prize for the development of ECR plasma etching technology.

Chapter 1
The Contribution of Dry Etching Technology to Progress in Semiconductor Integrated Circuits

Advances in the computer industry in recent years, as seen in the rapid commercialization of advanced information systems such as multimedia devices, is underpinned by various large-scale integration (LSI) devices such as microprocessors and memory. The LSI technology is advancing very rapidly, as shown in Fig. 1.1, with the device density doubling approximately every 2 years [1]. The transistor count in Fig. 1.1 refers to the number of transistors on each microprocessor chip, and this trend toward higher device density follows Moore's law. Minimum feature sizes are shrunk by approximately 30 % every 3 years, and devices with 20-nm-level minimum feature sizes are in volume production as of 2014.

A higher LSI device density means that there are a larger number of devices built into each chip, and the key is to make each device as small as possible. The fundamental technology for realizing this is fine geometry processing, which is mainly comprised of the lithography and dry etching technologies. Lithography is a technology for forming desired circuit patterns on the resist, which is a photosensitive material [2]. Dry etching is a technology for transferring the circuit patterns, formed with the resist, onto the underlying thin film by partially removing the various thin films deposited on a wafer, with this resist used as a mask. This chapter provides an overview of the dry etching technology and the roles this manufacturing process step plays in enhancing LSI device density.

Reducing manufacturing cost is also an important consideration with LSI manufacturing. Wafers with ever-larger diameters have been used, as shown in Fig. 1.2, to achieve this objective. As shown in Fig. 1.3, a large number of LSI chips are formed on a wafer. Their chip patterns are created on the wafer by steppers, in a step-and-repeat lithography process [2]. When larger wafers are utilized in the manufacturing process, the cost of each chip goes down because more chips are obtained from each wafer. The largest wafer diameter in manufacturing today is 300 mm, with 450-mm wafers planned next. Currently, 450-mm Si wafers are ready for manufacturing, and 450-mm semiconductor equipment, including dry etching equipment, is under development. As explained thus far, device scaling and larger

K. Nojiri, *Dry Etching Technology for Semiconductors*,
DOI 10.1007/978-3-319-10295-5_1

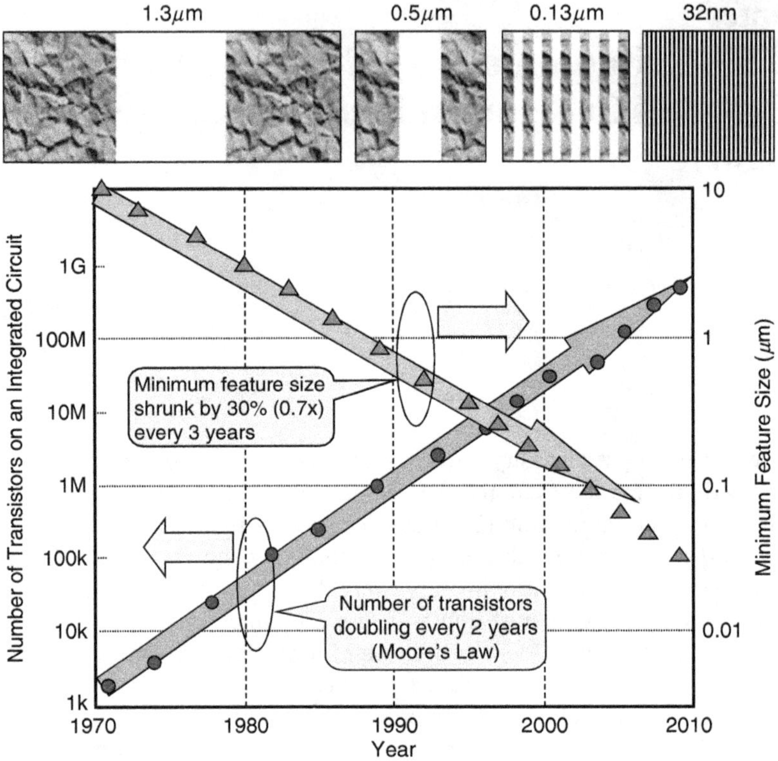

Fig. 1.1 Trends in the miniaturization and integration density of the semiconductor integrated circuit [1]

Fig. 1.2 Trends in wafer size increase

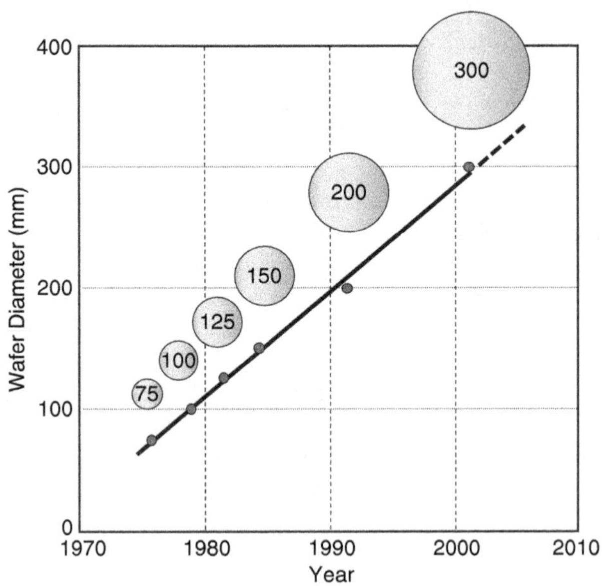

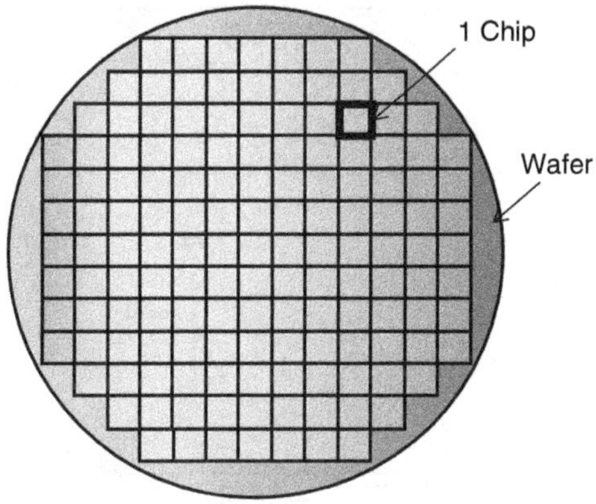

Fig. 1.3 LSI chips formed on a wafer

wafer diameters are indispensable in the semiconductor business. Technology innovations in dry etching are needed to achieve both.

1.1 Dry Etching Overview

Figure 1.4 shows an overview of dry etching. We'll discuss a parallel-plate dry etching system as an example. This type of etching equipment is called a reactive-ion etching (RIE). After the etch chamber is first pumped down to high vacuum, an etching gas is introduced. A plasma is generated next, when a 13.56-MHz radiofrequency (RF) power is applied to a pair of electrodes that face each other. The etch gas is dissociated in this plasma to generate reactive species, such as ions and radicals, and monomers, which form the basis of polymers. These reactive species and monomers are transported onto the wafer surface and react with the etch target material. As shown on the left-hand side in Fig. 1.4, complicated competitive reactions of etching and deposition take place near the wafer. In this example, CF and CF_2 monomers form polymers and deposit on the pattern surface. These polymers interact with ions and F radicals and are stripped off in the form of CF_4, while the underlying Si is etched through the interactions with F and Cl radicals. The byproducts of these reactions desorb from the wafer surface, and the etching proceeds. The etch byproducts ultimately go through the exhaust gas treatment system and are released into the atmosphere.

Figure 1.5 shows a dry etching process flow. A gate etching process is shown as an example; the steps involved in this process follow:

1. A gate oxide film is first formed on Si substrate. Then poly–Si, which is the gate material, is deposited on top.
2. Next, gate patterns are formed on a resist mask by lithography technology.

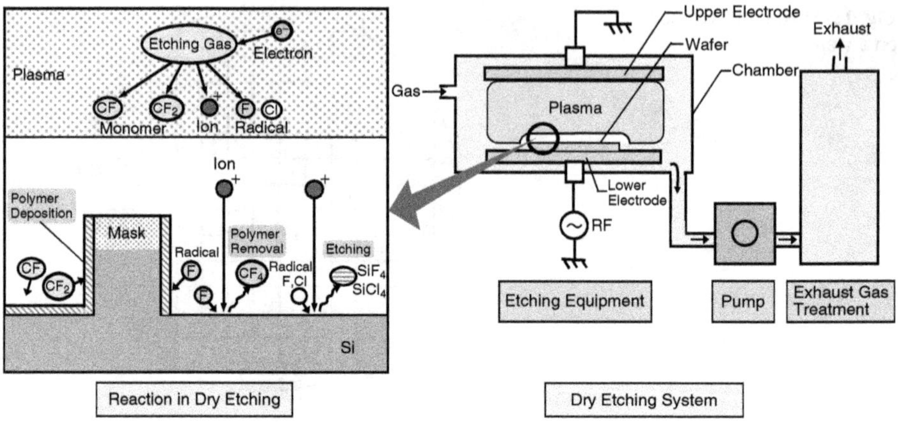

Fig. 1.4 Outline of dry etching

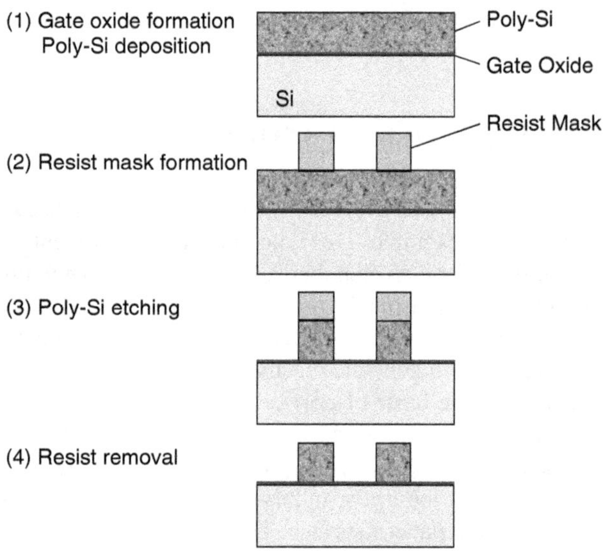

Fig. 1.5 Process flow for dry etching

3. The wafer is placed in the dry etching equipment. The underlying poly–Si is etched, with the resist used as the mask, and the etching stops when the gate oxide film is exposed.
4. The resist, which is no longer needed, is then stripped off, and the poly–Si gate etching is completed.

Figure 1.6 shows an SEM photograph of a 32-nm gate etching profile [3].

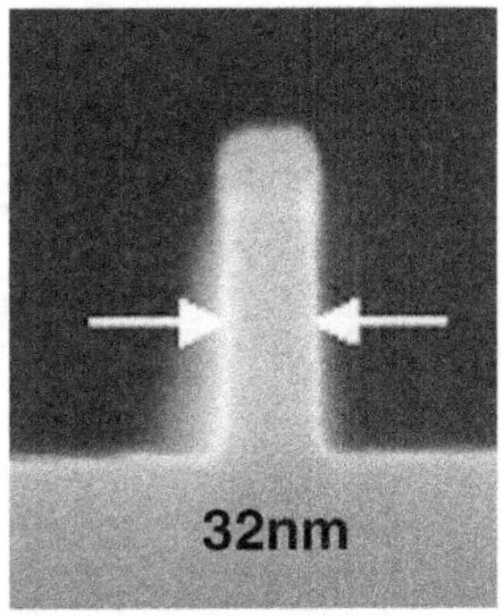

Fig. 1.6 Examples of the etched profile of a 32-nm gate [3]

1.2 Parameters for Evaluating Dry Etching Performance

The parameters required for evaluating the performance of dry etching are explained in Fig. 1.7. The first parameter is the etch rate. The etch rate (ER1) of an etch target film is calculated based on the etch target film thickness and the time to remove it completely. ER1 is directly related to the throughput, and process conditions should be chosen to make it as high as possible. The etch rate (ER2) of the underlying film and the etch rate (ER3) of the mask are calculated from the etch amount of each film.

A ratio of etch rates (ER1/ER2) between the etch target film and the underlying film and a ratio of etch rates (ER1/ER3) between the etch target film and the mask are called the selectivity to the underlying film and the selectivity to the mask, respectively. These are parameters that indicate how much the underlying film and mask are etched as the etch target film is etched. With dry etching, overetching is always needed, because there are always etch residues at the end point due to the etch rate nonuniformity and the underlying layer topography. A higher selectivity means less underlying film is etched away during overetching, and a higher selectivity is required as devices continue to be scaled. With the gate process, for example, the gate oxide film, which is the underlying film, becomes thinner with scaling. The gate oxide film could easily be etched through without a high selectivity, and the LSI yields would suffer as a result. For this reason, a high selectivity is required against the underlying gate oxide film. The same would be true with the selectivity to the mask. Thinner resist films are used to enhance the lithography resolution on highly scaled devices. A high selectivity to the resist is necessary for ensuring that the resist is not stripped off during etching.

The critical dimension (CD) is the finished dimension after the etching is complete. The CD is the most important parameter in the fine geometry processing technology. The CD, for example, has a direct impact on transistor performances with the gate process. In other words, the CD determines the metal oxide semiconductor (MOS) transistor threshold voltage V_{th}. When the CD is nonuniform across a wafer, V_{th} also becomes nonuniform across the wafer and causes yield losses. Therefore, etching with better uniformity increasingly becomes necessary. MOS transistors are discussed in more detail in the next section.

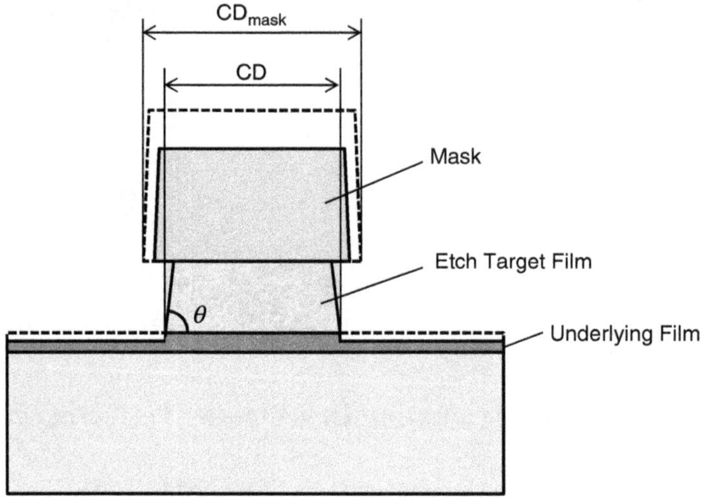

Parameters for evaluation

1. Etch rate : Etch target film (ER_1), Underlying film (ER_2), Mask (ER_3)
2. Selectivity : Etch target film/Underlying film=ER_1/ER_2, Etch target film/Mask=ER_1/ER_3
3. Dimension after etching : CD
4. CD shift : Δ CD= Mask CD (CD_{mask})−CD
5. Etch profile : Taper angle (θ)

Fig. 1.7 Parameters for evaluating dry etching performance

An amount of deviation from the critical dimension on the mask (CD_{mask}) is called CD bias (ΔCD), defined as $\Delta CD = CD_{mask} - CD$.

The final parameter is the etch profile. Ideally, the angle of taper θ should be 90° in a vertical etch profile. Reverse tapering, with θ being greater than 90°, must be avoided, because it creates shadows during ion implantation and adversely affects the transistor performances. Process conditions must be set to avoid reverse tapering.

1.3 Role of Dry Etching Technology in Miniaturization and Device Density Increase in LSI

This section discusses where dry etching technology is used in the LSI manufacturing process. Let's take the example of manufacturing a dynamic random access memory (DRAM) device. Figure 1.8 shows a cross section of a typical DRAM memory cell. A DRAM memory cell consists of one MOS transistor and one capacitor. A circuit diagram is shown on the right-hand side of Fig. 1.8. Electrons passing through the MOS transistor are stored in the capacitor. Data are read as either a

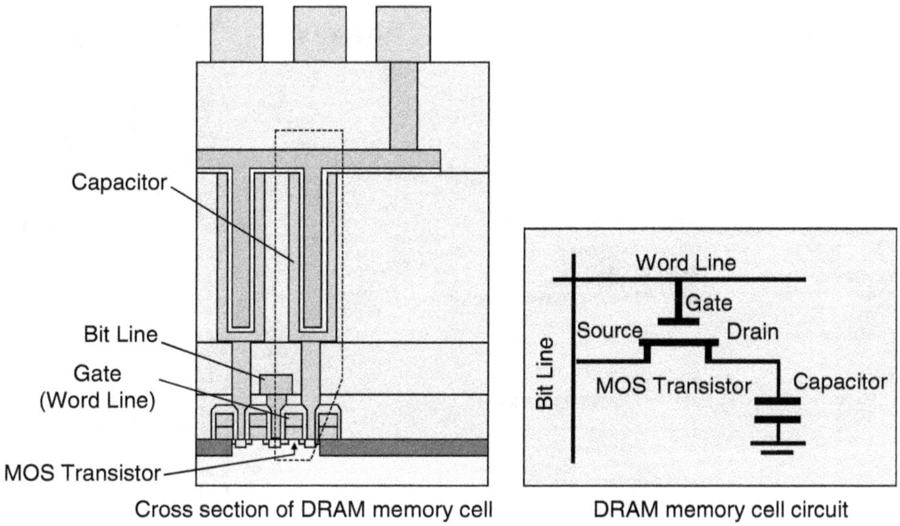

Fig. 1.8 DRAM memory cell structure

"1" or "0," depending on whether or not the capacitor is charged, respectively. Figure 1.9 shows a manufacturing process flow for DRAM. Details of the process flow are as follows:

1. The DRAM manufacturing process starts from a bare Si substrate.
2. A shallow trench isolation (STI) is formed by etching the Si substrate. This step is called STI etching. The figure does not show the steps for forming the resist mask and removing it after STI etching. Also, the steps for forming and removing the resist mask have not been included in the remainder of this description.
3. Next, the STI trenches are filled with an insulating film, which is planarized by chemical mechanical polishing (CMP). Isolation regions that isolate the devices are thus formed. A gate oxide film is then formed.
4. Poly–Si, WSi_2, and SiN are deposited on top of the gate oxide film. An SiN/WSi_2/poly–Si gate structure is formed by gate etching. Then the source drain extension (SDE) region is formed by ion implantation.
5. Next, SiN is deposited on top of the gate, followed by anisotropic dry etching. Because the SiN film is thick in the vertical direction, SiN remains on the gate sidewalls without being etched off and forms the spacers. This step is called spacer etching. Then, a source drain (SD) implantation follows, and the source drain regions are formed. The MOS transistors are now formed. MOS refers to the metal (a gate metal material, which is WSi_2/poly–Si here), oxide (the gate oxide film here), and semiconductor (Si substrate here). A current flowing between the source and drain of an MOS transistor is controlled through a voltage applied on the gate. In this book, all "transistors" refer to "MOS transistors."

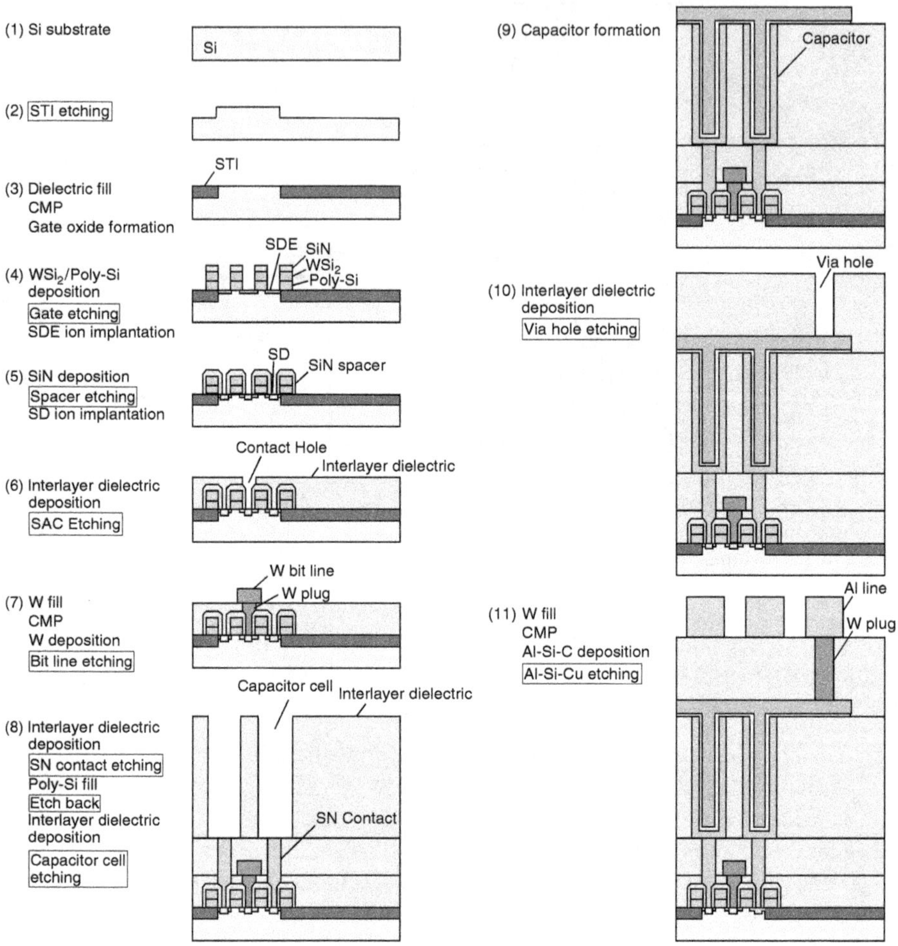

Fig. 1.9 Manufacturing process flow for DRAM

6. After an interlayer insulating film is deposited, contact holes to the Si substrate are opened by self-aligned contact (SAC) etching. Since the gates are covered with an Si_3N_4 film, which is an etch stop layer, the contact holes and the gate would not be shorted even when there is a misalignment. This technology is described in more detail in Chap. 3.
7. Next, contact holes are filled with tungsten (W) and planarized by CMP. They are called the tungsten plug. Tungsten is then deposited on top and etched to form the bit lines.
8. An interlayer insulating film is deposited, followed by storage node (SN) contact etching. The SAC etching process is also used here. Storage node contacts are filled with poly–Si and etched back for planarization. Next, a thick interlayer insulating film is deposited, followed by cell etching. Cell etching is a hole etching step for forming the capacitors and is tremendously challenging due to the very high aspect ratio (depth/hole diameter). High-aspect-ratio hole etching is described in detail in Chap. 3.
9. The cells are filled with a capacitor lower electrode (normally poly–Si), a capacitor insulating film, and a capacitor upper electrode (normally TiN), and the capacitors are thus formed.
10. After an interlayer insulating film is deposited, via holes for connecting the capacitors and metal lines are etched.
11. Via holes are filled with W and planarized by CMP. Next, Al–Si–Cu is deposited and etched, and Al metal lines are thus formed. The DRAM memory cell is now completed.

As thus described, an LSI device is created through repeated deposition and dry etching steps. Therefore, the etch processing accuracy will greatly determine the device performances and yields. While Fig. 1.9 only shows the simplified DRAM fabrication steps, it nevertheless illustrates the important roles of the dry etching technology in the LSI manufacturing process. Each dry etching process mentioned in this process flow is described in more detail in Chap. 3.

References

1. Y. Suzuki, T. Yunogami: The Doshisha Business Review, Vol. 60, No. 3-4, p.54 (2008).
2. S. Okazaki, A. Suzuki, T. Ueno: Lithography Technology for Semiconductors, Beginner's Books Series 29, Kogyo Chosakai Publishing Co., Ltd. (2003).
3. S. Ramalingam, Q. Zhong, Y. Yamaguchi, and C. Lee: Proc. Symp. Dry Process, p139 (2004).

Chapter 2
Mechanism of Dry Etching

A guiding principle for designing a dry etching process has yet to be established. However, guidelines for designing a process may be obtained by examining the reaction processes. For that, one must first understand the mechanism of dry etching. This chapter starts with the basics of plasma and goes on to describe the dry etching reaction processes and the mechanism of anisotropic etching without relying on mathematical equations or difficult theories, in a way that is completely accessible to readers who have no background in dry etching.

In the discussion on the fundamentals of plasma, all the topics for understanding dry etching mechanisms are covered, including "What is a plasma?", the plasma parameters, and the collisions and reactions that take place in the plasma.

Then there is a discussion on the behaviors of ions, which play an important role in realizing an anisotropic etching. Specific numerical values associated with the ion sheath thickness and mean free path are provided to help readers deepen their understanding of the ion scattering phenomena in the ion sheath. Next, a discussion on the dry etching reaction processes, mechanisms of anisotropic etching, and parameters that determine the etch rates and selectivity provides guidelines on how to select gases, set pressures, and build the process.

2.1 Basics of Plasma

2.1.1 What Is a Plasma?

This section begins with a simple description of plasma. Plasma refers to an "ionized gas," in which approximately the same numbers of electrons and ions exist; it is in an electrically neutral state from a macroscopic viewpoint. The electron density (n_e) and ion density (n_i) are substantially equal to one another, and they are referred to as the plasma density. Because electrons are able to travel freely

K. Nojiri, *Dry Etching Technology for Semiconductors*,
DOI 10.1007/978-3-319-10295-5_2

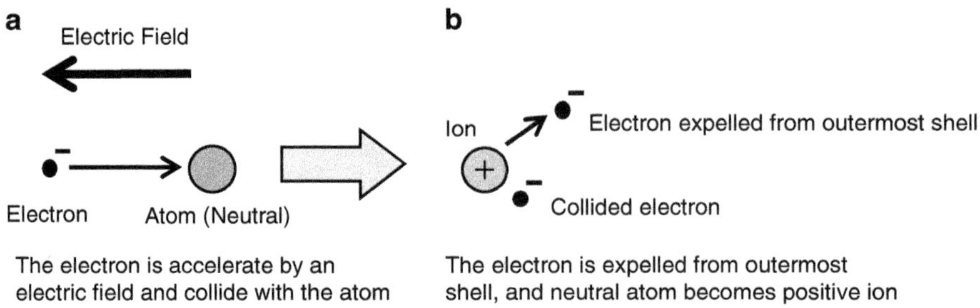

Fig. 2.1 Ion generation by the collision of electrons and neutral atoms

Fig. 2.2 Principle of gas discharge

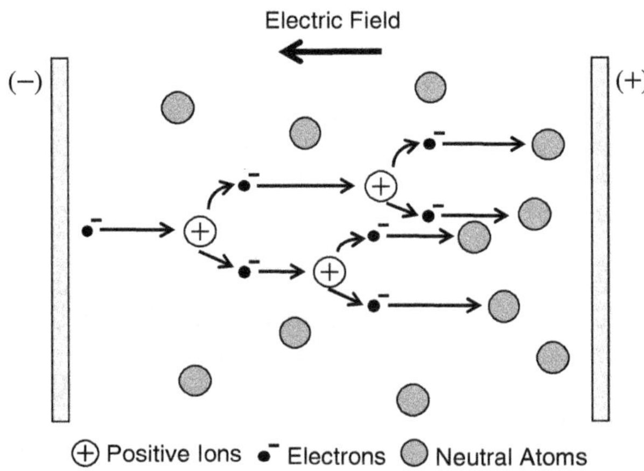

within the plasma, it has a conductive property. When a radiofrequency (RF) power is applied on a pair of electrodes in an etch chamber, electrons are accelerated by an electric field generated by the RF power, acquire kinetic energy, and collide with atoms and molecules (Fig. 2.1a). If the kinetic energy of an electron is greater than the ionization energy (ionization voltage), the electron in the outermost shell of the atom or molecule is expelled. As a result, the neutral atom or the molecule turns into an ion (Fig. 2.1b). On the other hand, the electron that has been expelled from the molecule or the atom adds to the first colliding electron to now make a total of two electrons. These electrons are then accelerated under the electric field, collide with other atoms and molecules, and generate new ions and electrons. The number of ions and electrons increase as in an avalanche and eventually exceed a threshold level over which a resulting discharge begins and creates a plasma. This mechanism is summarized in Fig. 2.2.

The plasma is classified into a completely ionized plasma, in which 100% of electrons and ions are ionized, and a weakly ionized plasma, in which the degree of ionization is low and a mixture of ions, electrons, and neutral atoms and molecules coexist. A glow discharge is used for dry etching. A plasma generated by the glow

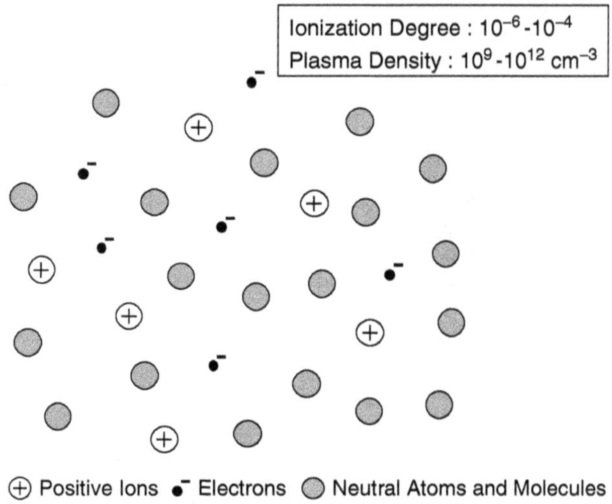

Fig. 2.3 Schematic of glow discharge plasma (weakly ionized plasma)

discharge is categorized as a weakly ionized plasma and consists of equal numbers of positive and negative charges, as well as electrically neutral atoms and molecules. Figure 2.3 shows a model diagram for a glow discharge plasma. The degree of ionization in a glow discharge plasma is on the order of 10^{-6}–10^{-4}. In other words, the degree of ionization is around 1 out of every 10,000 at most. The majority of the particles remain neutral, and only one ion and one electron exist for every 10,000 neutral particles. This is why it is called a weakly ionized plasma. The number of gas molecules at a pressure of 13.3 Pa (100 mTorr) is around 3.5×10^{15} cm^{-3}, so the plasma density at a ionization degree of 10^{-4} is 3.5×10^{11} cm^{-3}. The plasma density of a glow discharge plasma is within a range of 10^{-9}–10^{-12} cm^{-3}.

2.1.2 Plasma Parameters

Table 2.1 shows typical values for the plasma parameters associated with an arc discharge plasma, which is a strongly ionized plasma, and a glow discharge plasma, which is a weakly ionized plasma [1, 2]. A glow discharge plasma is characterized by a lack of thermal equilibrium between the electron temperature T_e and gas temperature T_g. An electron temperature corresponds to the energy of the electrons, and its relationship to the kinetic energy $\frac{1}{2}m_e v_e^2$ is expressed as

$$\frac{1}{2}m_e v_e^2 = \frac{3}{2}kT_e \tag{2.1}$$

where m_e is the electron mass, v_e is the electron velocity, and k is Boltzmann's constant.

Table 2.1 Types of plasma and plasma parameters [1, 2]

Type of plasma		Plasma density (cm^{-3})	Electron temperature T_e (K)	Ion temperature T_i (K)	Gas temperature T_g (K)
Arc discharge	Strongly ionized plasma (high-temperature plasma)	$>10^{14}$	6,000	6,000	6,000
Glow discharge	Weakly ionized plasma (low-temperature plasma)	10^9–10^{12}	$\sim 10^4$	300–1,000	300

Because electrons are very light, they are accelerated by the electrical field and acquire a large kinetic energy. The average electron energy in a glow discharge plasma is several electron-volts. Say the electron energy is 2 eV; then the electron temperature T_e is 23,200 K, according to Eq. (2.1). On the other hand, the temperature of the neutral atoms and molecules, which is the gas temperature T_g, is around room temperature (293 K). In other words, $\frac{T_e}{T_g}$ is around 80, and the electron temperature T_e and gas temperature T_g are not in a thermal equilibrium. Although the electrons have an energy level comparable to a high temperature of 10^4 K or higher, the etch chamber and the wafer remain at low temperature because the electron mass is small. For this reason, a glow discharge plasma is also referred to as a low-temperature plasma. Because electrons have enough energy for causing the excitation, ionization, and dissociation of atoms and molecules, with the gas temperature remaining at close to the room temperature, diverse types of reactions are possible at low temperature. This is the reason why the glow discharge plasma is used for semiconductor manufacturing.

An arc discharge is a strongly ionized plasma, and its plasma density is 10^{14} cm^{-3} or greater. The electron temperature T_e, ion temperature T_i, and gas temperature T_g are in a thermal equilibrium, and $T_e = T_i = T_g$ is approximately 6,000 K. For this reason, the arc discharge is referred to as a high-temperature plasma.

2.1.3 Collision Processes in a Plasma

Electrons that have gained energy in a plasma collide with atoms and molecules. These collisions are categorized as elastic collisions and inelastic collisions. Figure 2.4 summarizes the collision processes in a plasma. With an elastic collision, only the kinetic energy changes; the internal energy does not change. This type of collision tends to take place when the electron energy is low. In the example in Fig. 2.4, the electron is bounced back in a different direction. Because a portion of the electron's energy is transferred into the kinetic energy of the atom, the atom slightly gains a velocity. The electron loses a small amount of energy

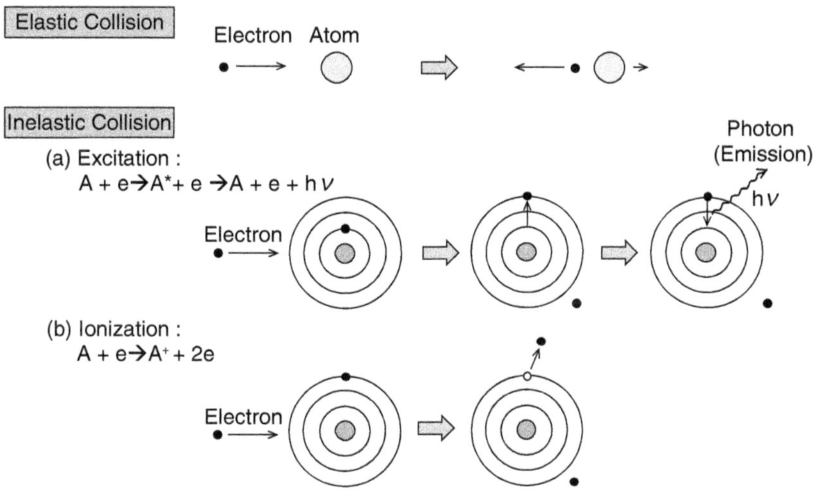

Fig. 2.4 Collision processes in a plasma

through the collision. With an inelastic collision, the internal energies are converted, and excitation, ionization, dissociation, and electron attachment take place, described next.

1. Excitation: A colliding electron provides energy to the bound electron in an atom and enables it to jump to a higher energy level. In general, the excited state is unstable, and the excited electron would be able to remain in this state for only around 10^{-8} s, and then returns to the ground state. Photon is emitted during this transition. The plasma glows because of this principle. An excitation reaction step is described as follows:

$$A + e \rightarrow A^* + e \rightarrow A + e + hv$$

   where A represents a neutral atom, and A* represents A in an excited state. h is Planck's constant, and ν is the frequency of the emitted light.
2. Ionization: As mentioned earlier, an electron in the outermost shell is expelled when the energy of the colliding electron is larger than the ionization voltage, and the neutral atom turns into a positive ion. This reaction step is described as follows:

$$A + e \rightarrow A^+ + 2e$$

3. Dissociation: Dissociation occurs when the energy given by the colliding electron is larger than the binding energy of the molecule. This reaction step is described as follows:

$$AB + e \rightarrow A + B + e$$

   When a molecule is dissociated, its byproducts are chemically more active than the original molecule and turn into highly reactive particles. A particle in this activated state is called a radical. It has been reported that CF_4 would easily be dissociated into a CF_3 radical ($CF_3\bullet$) and F radical ($F\bullet$) once excited [3]. This reaction step is described as follows:

$$CF_4 \rightarrow CF_4{}^* \rightarrow CF_3\bullet + F\bullet$$

4. Electron attachment: The colliding electron attaches to the atom and turns it into a negative ion. This reaction step is described as follows:

$$A + e = A^-$$

2.2 Ion Sheath and Ion Motion in the Sheath

2.2.1 *Ion Sheath and* V_{dc}

V_{dc} is an extremely important parameter for understanding the dry etching mechanism and is thus described first here. In the parallel-plate dry etching equipment [reactive-ion etching (RIE)] shown in Fig. 2.5, a lower electrode, on which a wafer is placed, is connected to an RF power supply through a blocking capacitor, and the opposite electrode (upper electrode) is connected to the ground. A frequently used RF frequency is 13.56 MHz; in other words, the direction of the electric field changes 13.56×10^6 times every second. Because electrons have a small mass, they are able to follow the oscillation of the electric field. On the other hand, ions are much heavier, with masses that are approximately 100,000 times larger, and they are unable to follow the RF oscillation and do not move much from where they are. As a result, only the electrons accelerated by the electric field jump into the electrodes. Because the lower electrode is connected to a blocking capacitor, it is gradually biased to a negative potential. A direct current (DC) bias that is thus generated is referred to as a self-bias and is represented by V_{dc}. While the V_{dc} value is dependent on the RF power, it would be in the tens of volts to several hundreds volts for

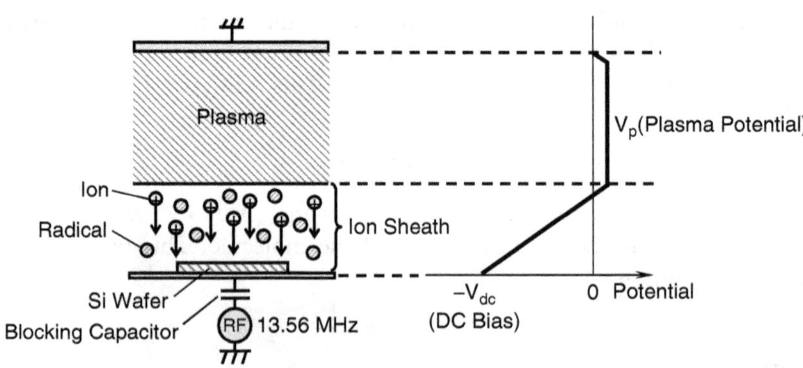

Fig. 2.5 Ion sheath and V_{dc}

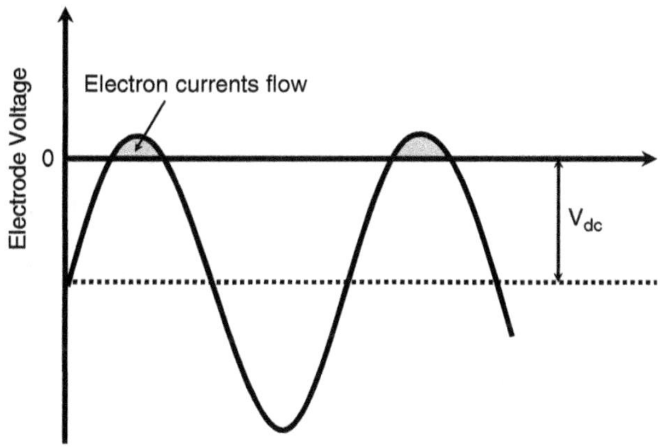

Fig. 2.6 Voltage waveform at the lower electrode in an RF discharge

the etching of conductive materials such as Si and Al. Figure 2.6 shows the voltage waveform at the lower electrode in a steady state. The electrode voltage turns positive only for a small period of time during each cycle. Electron currents flow onto the electrode during this period. On the other hand, ion currents flow almost continuously. The sum of the amounts of injected charges in each cycle is zero in a steady state.

Electrons are pushed away when the electrode is negatively biased, and almost no electrons exist near the electrode. This region is called the ion sheath, with the analogy being the sheath of a sword. In other words, an ion sheath refers to a sheath filled with ions. Because the electron density is very low in this region, the probability of collision excitation is low, and almost no light is emitted. For this reason, an ion sheath is also referred to as a dark space.

Figure 2.5 shows an electrical potential distribution inside the etch chamber. As mentioned earlier, the plasma is electrically conductive, and the plasma is at equipotential from a macroscopic viewpoint; for this reason, ions travel in random directions in the plasma. The electrical potential of the plasma is referred to as the plasma potential V_p. On the other hand, ions approaching the interface between the plasma and the ion sheath are accelerated toward the wafer due to the $-V_{dc}$, which creates a potential gradient inside the ion sheath and toward the lower electrode. Here, the ions acquire the energy equivalent to $V_p + V_{dc}$. Etching is driven mainly by these ions, which travel in a straight direction, and an anisotropic profile, with small dimensional shifts from the mask patterns, is obtained. This mechanism is described in detail in the next section.

The magnitude of voltage induced at the electrodes is dependent on the ratio of the electrode surface areas. Say that the surface areas of electrodes 1 and 2 are S_1 and S_2, respectively, and the voltages induced at each electrode are V_1 and V_2, respectively, as shown in Fig. 2.7. Their relationship [4] would be

$$\frac{V_1}{V_2} = \left(\frac{S_2}{S_1}\right)^4 \tag{2.2}$$

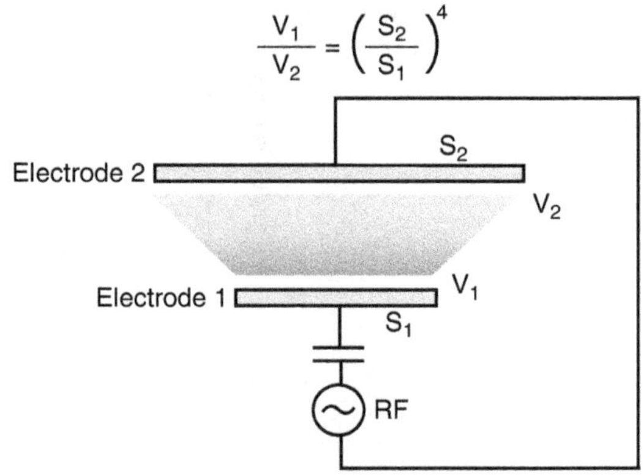

Fig. 2.7 Relationship between the electrode area ratio and the voltages induced at electrodes

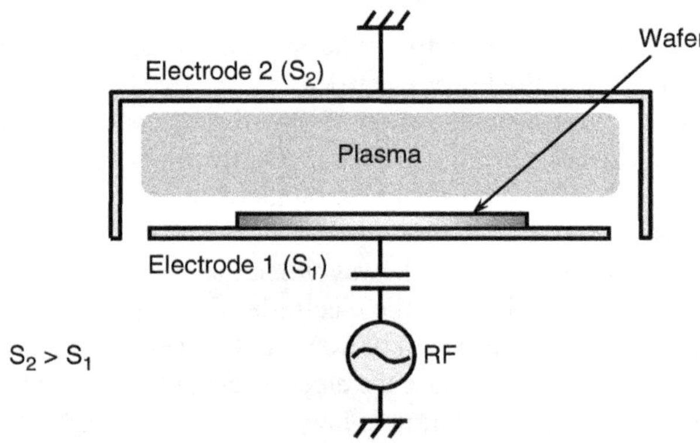

Fig. 2.8 Electrode area ratio in RIE system

In other words, a higher voltage is induced at the electrode with a smaller surface area. In general, the electrode (electrode 2) that opposes the electrode (electrode 1) on which the wafer is placed has a much larger surface area in the RIE system, so that an adequate V_{dc} is achieved at the wafer. As shown in Fig. 2.8, the chamber walls are normally held at the same electrical potential as electrode 2 in the RIE system, so that the structure helps make S_2 larger than S_1. Because the V_{dc} induced at the chamber walls is small in this instance, sputtering on the walls is suppressed, and as a result there is an added advantage of suppressing the release of impurities, such as heavy metals from the chamber walls.

2.2.2 *Ion Scattering in the Sheath*

Let's now quantitatively examine the scattering of ions in the sheath. The ion sheath thickness (d_{is}) is represented by the following Child–Langmuir equation [5]:

$$d_{is} = \frac{2}{3}\left(\frac{\varepsilon_0}{i_{io}}\right)^{\frac{1}{2}}\left(\frac{2e}{m_i}\right)^{\frac{1}{4}}\left(V_p - V_{dc}\right)^{\frac{3}{4}} \qquad (2.3)$$

Table 2.2 Calculated sheath thickness and related constants of high-density plasma [6]

Conditions	
Gas	Ar
Gas pressure	1.33 Pa
Mean free path (λ)	5 mm
$V_{dc}-V_p$	−100 V
Ion current density	15 mA/cm^2
Calculated results	
Ion sheath thickness (d_{is})	0.28 mm

where i_{io} is the ion current density, ε_0 is the permittivity of vacuum, e is the elementary electric charge, m_i is the ion mass, and V_p is the plasma potential.

Table 2.2 shows an example of the ion sheath thickness calculated for a high-density plasma [6]. The ion sheath thickness would pretty much be at this value for high-density plasmas such as the inductively coupled plasma (ICP) and the electron-cyclotron resonance (ECR) plasma, which are described in Chap. 4. In this example, the ion sheath thickness is calculated for Ar at a pressure of 1.33 Pa (10 mTorr) and an ion current density of 15 mA/cm^2. The ion sheath thickness d_{is} is extremely small, at 0.28 mm.

Another important concept in examining the scattering of ions in the sheath is the mean free path (λ), which we briefly discuss here. A mean free path is an average distance that a particle travels from one collision to the next. In other words, it is the distance over which a particle is able to travel without a collision. The mean free path will be longer at a lower gas pressure because fewer molecules exist. In other words, the lower the pressure, the longer the particle will be able to travel without a collision. The mean free path is inversely proportional to pressure. In the example shown in Table 2.2, the mean free path of Ar is 5 mm at 1.33 Pa, and it becomes 50 mm when the pressure is reduced to 0.133 Pa.

The scattering of ions in an ion sheath is examined by comparing the ion sheath thickness with the mean free path. In other words, when the mean free path is adequately larger than the ion sheath thickness, ions go through almost no scattering upon their arrival at the wafer. This directionality of ions is an important factor for enabling the anisotropic etching to be discussed in the next section.

We'll now compare the scattering of ions in an ion sheath with high-density plasma etching (e.g., ICP and ECR) and with the batch RIE (Fig. 2.9). The ion current density in a high-density plasma etcher is one order of magnitude larger than in the batch RIE. This high-density plasma etcher offers high etch rates and at the same time has a smaller ion sheath thickness. With a high-density plasma such as ICP and ECR, the sheath thickness d_{is} is thin at 0.28 mm. On the other hand, the mean free path λ is long at 5 mm because of the lower operating pressure of 1.33 Pa. In other words, λ is much larger compared to $d_{is}\left(\frac{d_{is}}{\lambda}=0.056\right)$, and the majority of ions arrive at the surface of the sample without colliding with neutral particles in the sheath (Fig. 2.9a). On the other hand, λ is smaller than d_{is} in the batch RIE $\left(\frac{d_{is}}{\lambda}=3.8\right)$, and most of the ions collide with neutral particles in the sheath and are scattered (Fig. 2.9b).

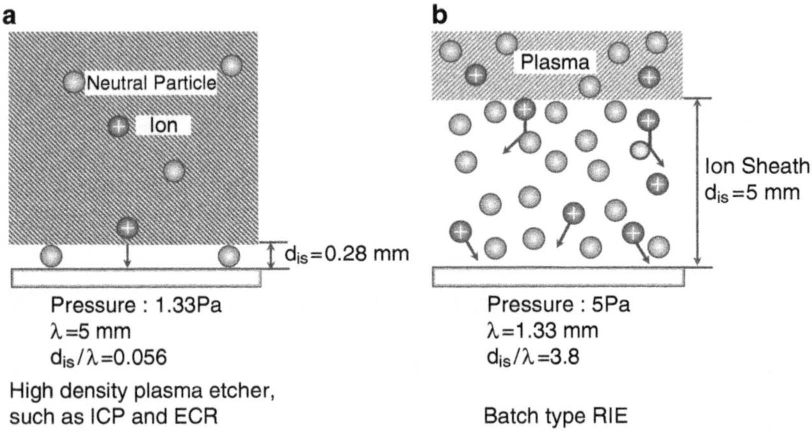

Fig. 2.9 Ion scattering in the ion sheath

Fig. 2.10 Effect of the ion direction on fine patterning

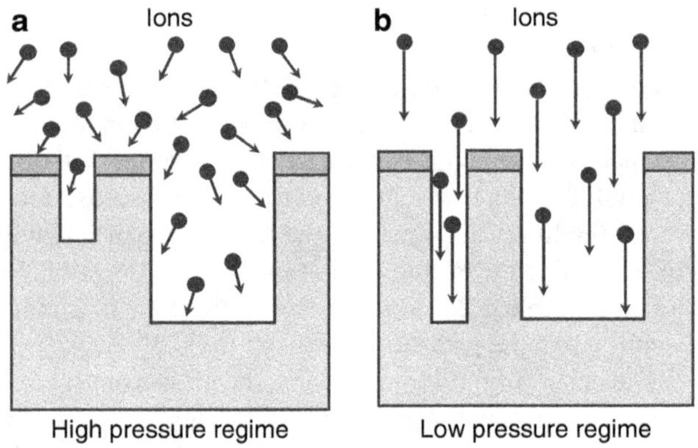

As thus explained, ion scattering in the sheath in the high-density plasma etcher is almost nonexistent, and there are very few ions coming in at an angle.

Figure 2.10 illustrates the effect of the ions' travel direction on fine patterning. When there are a large number of ions coming in at an angle due to scattering, fewer ions are able to enter fine patterns, as Fig. 2.10a shows, and the etch rate drops. On the other hand, an adequate number of ions are able to enter patterns with large openings, and the etch rate does not suffer. In other words, the etch rate becomes pattern-dependent. This issue can be resolved by using a lower-pressure regime, which prevents ions from being scattered and coming in at angles. As Fig. 2.10b shows, ions that are traveling highly directionally are able to etch fine patterns and wider patterns at the same etch rate. Figure 2.11 illustrates this phenomenon, based on the study of the various types of etchers [6]. In plasma etcher with a high operating pressure of 270 Pa, the etch rate of contact holes strongly depends on the hole size. The etch rate drastically decreases below a diameter of 1.0 μm. The decrease

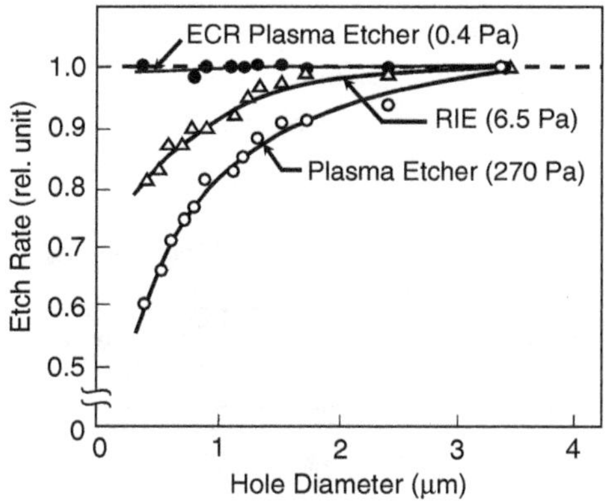

Fig. 2.11 Etch rates of contact holes as a function of hole diameter and operating pressure in three different etches [6]

in the etch rate is 40 % at a diameter of 0.4 μm. On the other hand, the ECR plasma etcher with a low operating pressure of 0.4 Pa shows a plateau in the etch rate characteristics down to a diameter of 0.4 μm, because there are enough incoming ions even in smaller holes.

2.3 How to Design Dry Etching Processes

2.3.1 Reaction Processes in Dry Etching

As mentioned at the beginning of this chapter, it is necessary first to examine the reaction processes in order to obtain the guidelines for designing a dry etching process. Let's now go over the basic concepts involved.

Dry etching proceeds in the following four steps: (1) Reactive species (neutral radicals and ions) are generated in the plasma; (2) species are transported and adsorb on the etch target film; (3) reactions take place at the surface of the etch target film, and etch byproducts are created; and (4) etch byproducts desorb from the surface of the etch target film. Figure 2.12 shows the reaction steps when Si is etched with CF_4 [7]. First, CF_4 dissociates into CF_3 radicals ($CF_3\bullet$) and F radicals ($F\bullet$) in the plasma.

Next, the CF_3 radicals adsorb on the Si surface and react with Si to form etch byproducts SiF_4. SiF_4 desorbs from the Si surface, and the etching thus proceeds.

In order to realize fine patterning, it is important to process with a high level of fidelity to the mask patterns. Figure 2.13 illustrates the differences between the isotropic and anisotropic etching. When radicals drive the etching reaction, the etching proceeds not only in the vertical direction, but also in the horizontal direction, as Fig. 2.13a shows, because the radicals move randomly by Brownian motion. This is referred to as isotropic etching, and it results in undercut under the mask and makes fine patterning difficult. On the other hand, as Fig. 2.13b shows, an etching that proceeds in a vertical direction would enable processing that closely replicates the mask patterns. This is referred to as anisotropic etching.

(1) Reaction species (neutral radicals and ions) are generated in the plasma
(2) Species are transported and adsorb on the target etch film
(3) Reaction takes place at the target etch film surface, and byproducts are created
(4) Byproducts desorb from the target etch film surface

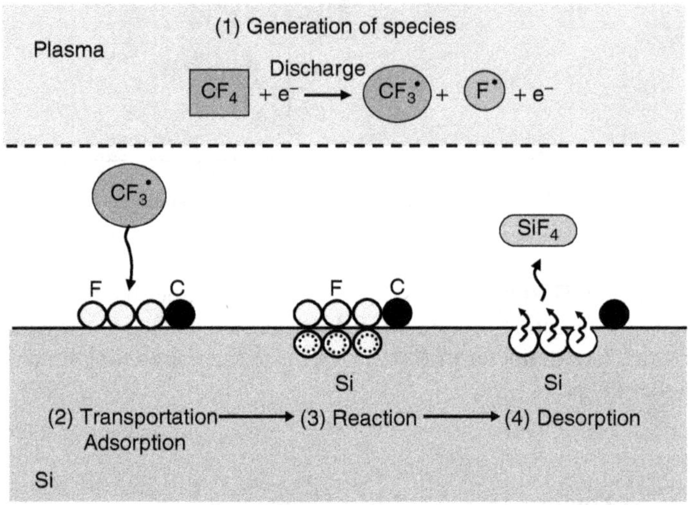

Fig. 2.12 Reaction steps of dry etching for the case of Si etching with CF_4 [7]

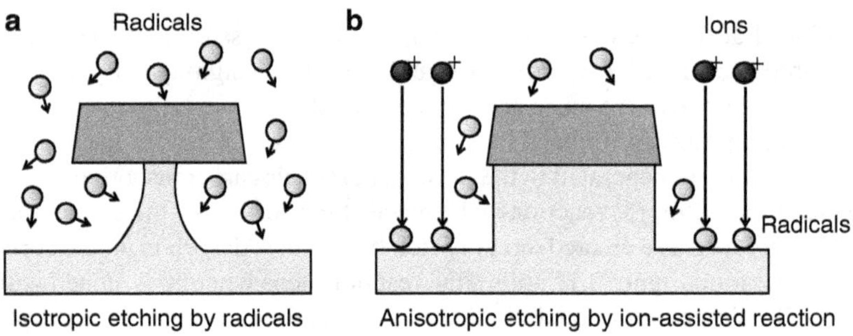

Fig. 2.13 Isotropic and anisotropic etching

2.3.2 Mechanism of Anisotropic Etching

Let's first consider how to achieve anisotropic etching. The basic approach to the implementation of anisotropic etching is to make the surface reaction proceed only in the vertical direction. Ion-assisted reactions are one approach to achieving this objective. An ion-assisted reaction is a phenomenon in which the incoming ions enhance the surface reactions. In a system in which ion-assisted reactions take

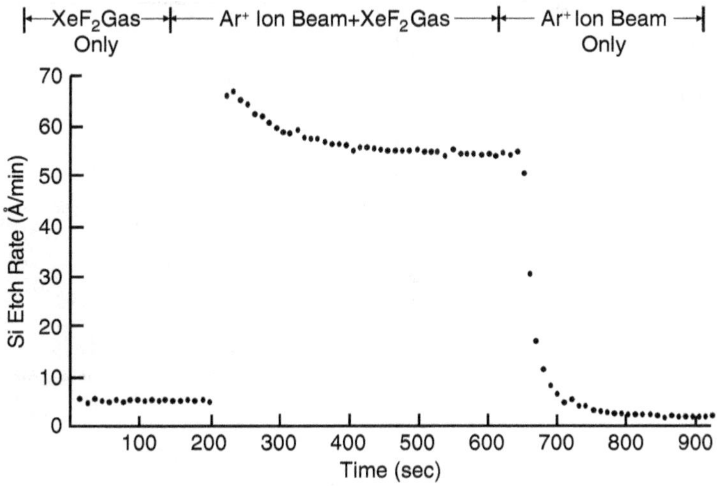

Fig. 2.14 Ion-assisted etching [8]

place, the etch rate of an ion-irradiated surface is significantly larger than the etch rate by neutral radicals. Therefore, an anisotropic etching is realized by making the ions strike the etch target surface at normal incidence.

Figure 2.14 shows the experimental result that explains the ion-assisted reactions [8]. The vertical axis in the figure represents the Si etch rate. First, Si is etched with only the XeF_2 gas. In this instance, Si is etched by the F radicals dissociating from XeF_2, but the etch rate is low, around 5 Å/min at most. The etch rate, on the other hand, increases by a factor of 10 or more when this reaction system is irradiated with Ar^+ ions at 450 eV. Next, when the supply of XeF_2 is halted, and only the Ar^+ ions are irradiated, the etch rate decreases drastically to 3 Å/min or less with only the physical sputtering by ions. This type of phenomenon is called the ion-assisted etching, in which the ion bombardment is given to the adsorbed radicals and the etch rate increases. The rate of etching would otherwise be extremely low when it is only by the radicals or by physical sputtering. There are several theories for the ion-assisted reaction. The most widely believed theory is the "hot spot model," in which the temperature of the areas under the ion irradiation becomes very high locally, and the etch rate goes up as a result of a significant boost to the radical reactions [9].

Because the rate of etching through the ion-assisted reactions is orders of magnitude larger than the rate of etching by radicals, as thus shown, it is possible to obtain an anisotropic etch profile when ions come in on the wafer vertically, so that the etch rate in the vertical direction (by assisted reactions) is faster than the rate of etching in the horizontal direction (by radicals). This is the principle of anisotropic etching by ion-assisted reactions.

Figure 2.15 shows the etch rate for each etching component as a function of the energy of impinging ions. Physical sputtering by ions is energy-dependent, and the etch rate increases with increasing ion energy. However, the etch rate itself is low.

Fig. 2.15 Etch rate for each etching component as a function of energy of impinging ions

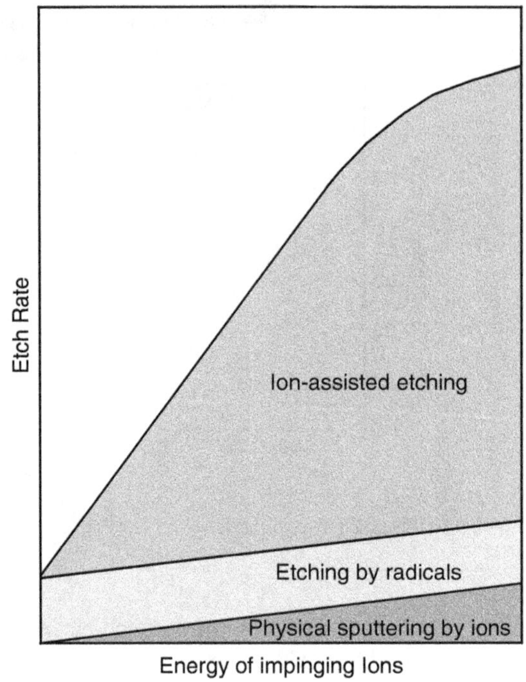

Chemical etching driven by radicals is not dependent on the energy, and so the rate always remains constant. The etch rate is also low. The etch rate of ion-assisted etching is extremely dependent on the ion energy, and the etch rate drastically increases with increasing ion energy. Anisotropic etching is realized through ion-assisted etching.

The effects of ion-assisted etching vary among various combinations of the etch target materials and etching gases. In other words, depending on the combinations of the etch target materials and etching gases, the ion-assisted reaction may or may not occur. Figure 2.16 shows the chemical sputtering yield of Si resulting from Cl^+, Br^+, and F^+ ion bombardment as a function of ion energy [10], and Fig. 2.17 shows the chemical sputtering yield of Al, B, C, and Si resulting from Cl^+ ion bombardment as a function of ion energy [9]. Here, the chemical sputtering yield is defined as a difference between the sputtering yield per ion and the physical sputtering yield [9]. According to Fig. 2.16, the chemical sputtering yield strongly depends on the ion energy when Si is etched with Cl^+ and Br^+. In other words, an ion-assisted reaction takes place with these combinations, and anisotropic etching is possible. On the other hand, there is a small dependence on the ion energy when Si is etched with F^+ ions. That is, the effect of the ion-assisted reaction is small, and anisotropic etching is less likely to take place. In the actual process, etch gases such as Cl_2 and HBr, which dissociate and generate the Cl^+ and Br^+ ions, are suited for anisotropic etching of Si. This is the reason why these gases are used as the base gases for the

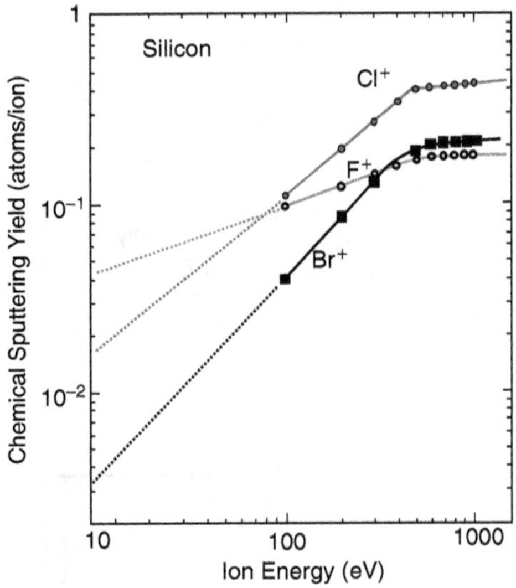

Fig. 2.16 Chemical sputtering yield of Si resulting from Cl^+, Br^+, and F^+ ion bombardment as a function of ion energy [10]

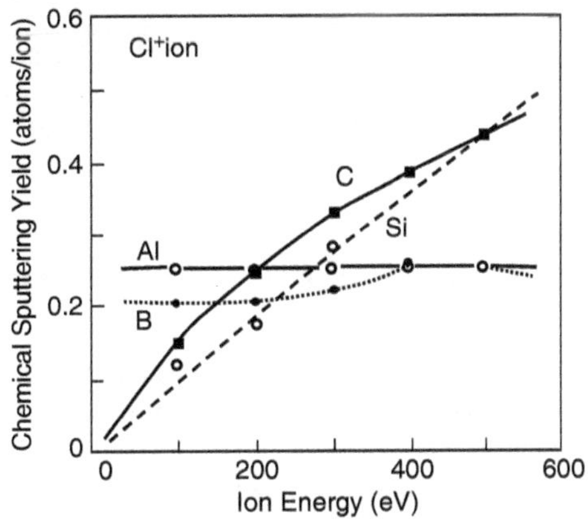

Fig. 2.17 Chemical sputtering yield of Al, B, C, and Si resulting from Cl^+ ion bombardment as a function of ion energy [9]

etching of poly–Si gates and shallow trench isolations (STIs). On the other hand, etching tends to become isotropic when Si is etched with gases like SF_6 and CF_4, which dissociate and generate the F^+ ions.

We'll now consider Al etching. According to Fig. 2.17, the chemical sputtering yield shows no dependence on the ion energy when Al is etched with Cl^+ ions. In other words, no ion-assisted reactions take place with this combination, and etching proceeds isotropically.

In order to achieve anisotropic etching with a system in which an ion-assisted reaction does not take place at all, or in a system where the effects of ion-assisted reactions are small, as thus shown, a process with sidewall protection is used.

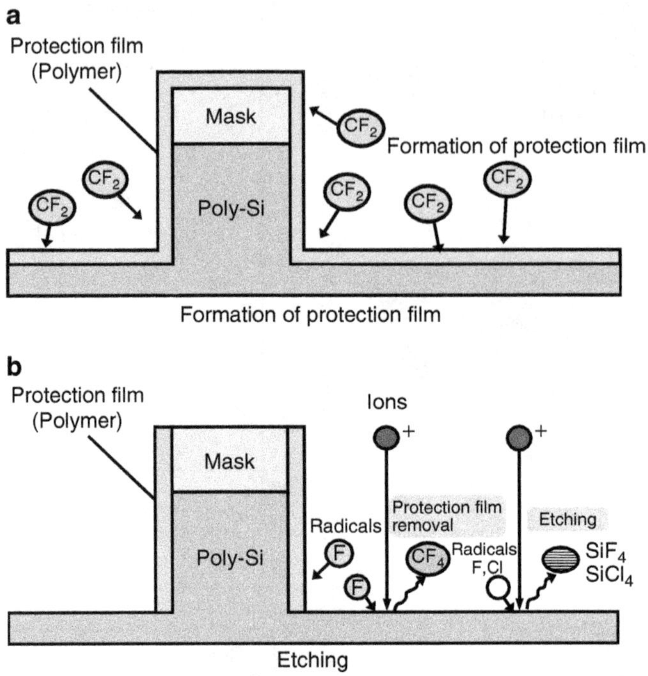

Fig. 2.18 Model for sidewall protection process [11]

2.3.3 *Sidewall Protection Process*

A sidewall protection process refers to a process in which the sidewalls of the etch target surface are protected with a protective film like polymer, and the protective film prevents invasion by radicals during etching. The protective film is formed when inorganic materials like SiO_2 or organic polymers are generated in the plasma. Figure 2.18 shows a model illustrating the sidewall protection process [11]. In this example, the protective film consists of organic polymers. In the case of using a mixture gas in which a gas for polymer formation is added to a main etch gas, monomers such as CF and CF_2 are generated from the additive gas. These monomers adsorb on the etch target surface and form the polymers.

As a result, the entire etch target surface is covered with polymers (Fig. 2.18a). Because ions come in vertically against the wafer, the ions remove the polymers on the horizontal surface, and the etching proceeds through the reactions between the exposed etch target material and the ions and radicals. On the other hand, because polymers are not stripped off and remain on the pattern sidewalls, where there are almost no incoming ions, the sidewalls remain protected by the polymers. Because the polymers prevent the attack of radicals on the sidewalls, an anisotropic etching is realized (Fig. 2.18b). This is the fundamental mechanism of the sidewall protection process. The reactions in parts (a) and (b) of Fig. 2.18 have been described separately here; they take place at the same time as the actual etching.

The monomers, which are the basis of polymer formation, are sometimes generated from the main etch gas itself. Furthermore, they may also be generated when the resist is etched. When Al is etched with a Cl-based gas chemistry, the etching should be isotropic in theory, because the ion-assisted reactions do not occur. In reality, however, an anisotropic profile is obtained. The reason for this is a supply of C and H resulting from the resist being etched, which forms the sidewall protection film.

As mentioned earlier, the effect of ion-assisted reactions is very small when Si is etched using an F-based gas like SF_6. In this case, SiO_2 would be generated when O_2 is added to the gas, and it forms a sidewall protection film. An anisotropic profile can then be obtained.

2.3.4 *Etch Rate*

The etch rate is related to all four of the reaction steps. In particular, surface reactions and desorption of etch byproducts are important considerations in selecting the etching gas. Some useful references are, first, the chemical sputtering yield, which was discussed earlier, and, second, the vapor pressures of etch byproducts. Figure 2.19 shows the vapor pressure curves in a simplified form. The figure shows that, in general, fluorides have higher vapor pressures than the chlorides. A higher vapor pressure means the etch byproduct is more volatile. In other words, when a gas chemistry that generates the fluorides is used, the etch byproducts desorb more

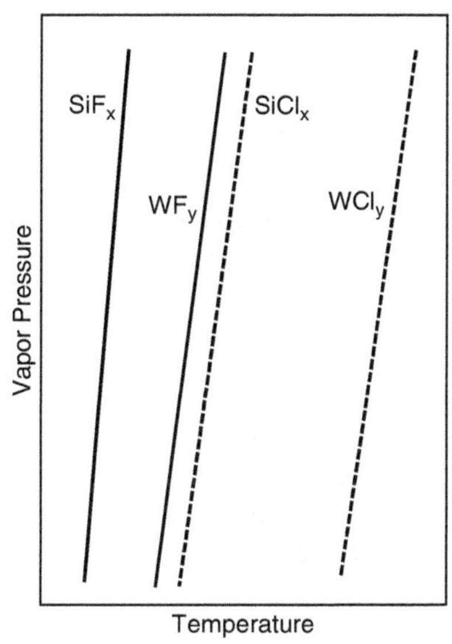

Fig. 2.19 Vapor pressure curves

Table 2.3 Melting and boiling points for various halides [12]

Halide	Melting point (°C)	Boiling point (°C)
SiF_4	–77 (2 atm)	–95 (sublimation)
$SiCl_4$	–70	57.6
$SiBr_4$	5.2	153.4
WF_6	2.5	17.5
WCl_6	275	346.7
AlF_3	1,290	–
$AlCl_3$	190 (2.5 atm)	183 (sublimation)
$AlBr_3$	97.5	255
CuF	908	1,100 (sublimation)
CuCl	452	1,367
CuBr	504	1,345

easily and the etch rate becomes higher. Data in Table 2.3 show the melting and boiling points for various halides [12]. They also show that the fluorides of Si and W tend to vaporize more easily than the chlorides. However, there is an opposite trend with Al, and the fluorides are less easy to vaporize than the chlorides. This indicates that it is possible to etch Al with a Cl-based gas, not with an F-based gas. Also, it should be noted that Al fluorides tend to result in particles, requiring caution. Table 2.4 shows the temperatures at which the various chemical compounds achieve a vapor pressure of 1,333 Pa (10 Torr) [13]. The pressure of 1,333 Pa has been chosen because it is assumed that a dry etching process would take place below this pressure. Based on this table, it is possible to figure out how easily each byproduct vaporizes and use it as a guideline for increasing the etch rates. The *Handbook of Chemistry* should be used as a reference when it is necessary to find the vapor pressures for etching the various materials [14].

2.3.5 *Selectivity*

For the selectivity to the underlying material, the ion energy and atom-to-atom bond strength should be considered [15]. Because the reactions proceed in a direction that results in a larger bond strength, it is possible to reduce the etch rate of the underlying material through the selection of gas chemistry based on the bond strength. The selectivity can then be improved. For example, gases based on Br and Cl, whose bond strength with Si is smaller than that of Si–O, might be used for the etching of poly–Si in order to increase the selectivity to the underlying SiO_2. Then the etch rate of the SiO_2 would be extremely low, and a high selectivity would be achieved [16]. This is described in detail in the section on poly–Si etching in the next chapter. Furthermore, another effective approach might be to lower the ion energy in a gas system in which the etch rate dependence on the ion energy is strong for SiO_2 and weak for poly–Si. An F-based gas is the case in point.

Table 2.4 Temperature at which the various chemical compounds achieve a vapor pressure of 1,333 Pa [13]

Chemical compound	Temperature (°C)	Chemical compound	Temperature (°C)	Chemical compound	Temperature (°C)	Chemical compound	Temperature (°C)
AgCl	1,074	CdI_2	512	HgI_2	204.5	$SbCl_3$	85.2
AgI	983	CrO_2Cl_2	13.8	KBr	982	SbI_3	223.5
$AlBr_3$	118.0	Cu_2Br_2	718	KCl	968	$SiCl_4$	−34.7
$AlCl_3$	123.8	Cu_2Cl_2	702	KF	1,039	$SiClF_3$	−127.0
AlF_3	1,324	Cu_2I_2	656	KI	887	$SiCl_2F_2$	−102.9
AlI_3	225.8	$FeCl_2$	700	LiBr	888	$SiCl_3F$	−68.3
As	437	$FeCl_3$	235.5	$MgCl_2$	930	SiF_4	−130.4
AsH_3	−124.7	$GeBr_4$	56.8	NaBr	952	$SnBr_4$	72.7
BBr_3	−10.1	$GeCl_4$	−15.0	NaF	1,240	$SnCl_2$	391
BCl_3	−66.9	H_2S	−116.40	NaI	903	$SnCl_4$	10.0
BF_3	−141.3	H_2S_2	−19	$NiCl_2$	759	SnH_4	−118.5
$CdCl_2$	656	H_2Se	−100	PbI_2	571	SnI_4	175.8
CdF_2	1,559	$HgBr_2$	179.8	PbF_2	904	$ZnCl_2$	508
		$HgCl_2$	180.2	$SbBr_3$	142.7	ZnF_2	1,359

2.3.6 *Summary*

As thus explained, when designing a dry etching process, the etching gas selection and reaction control should be chosen by considering the ion energy dependence of the chemical sputtering yield, the vapor pressures of the etch byproducts, and the atom-to-atom bond strength. In reality, however, it is difficult to satisfy the requirements associated with anisotropy, etch rate, and selectivity all at the same time with a single gas. For example, as Fig. 2.16 shows, Cl-based gases are suited for anisotropic etching of poly–Si. However, Fig. 2.19 shows that a high etch rate is difficult to achieve, because the vapor pressure of the etch byproduct $SiCl_4$ is lower than that of SiF_4. The opposite is true with the F-based gases. When designing an actual process, an appropriate mixture of these gases might be used, and a gas that would easily result in the formation of polymers might be mixed in for a sidewall protection process.

References

1. T. Iijima, S. Kondo, T. Aoyama: Plasma Technology, Beginner's Books Series 7, Kogyo Chosakai Publishing Co., Ltd. (1999).
2. Y. Hatta: Gas Discharge, 2nd Edition, Kindai Kagaku Sha Co., Ltd. (1971).
3. M. Tsuda: Plasma Process Technology for Semiconductor, Sangyo Tosho Publishing Co., Ltd., p.23 (1980).
4. H. R. Koenig and L. I. Maissel: IBM J. Res. & Dev. 14, p.168 (1970).
5. B. Chapman: Glow Discharge Processes, John Wiley & Sons (1980).
6. K. Nojiri and E. Iguchi: J. Vac. Sci. & Technol. B 13, 1451(1995).
7. H. Horiike: Proc. 19th Semiconductor Technology Seminar, p.193 (1981).
8. J. W. Coburn and H. F. Winters: J. Appl. Phys. 50, 3189 (1979).
9. S. Tachi: Proc. Symp. Dry Process, p.8 (1983).
10. S. Tachi and S. Okudaira: J. Vac. Sci. Technol. B 4, 459 (1986).
11. K. Nojiri, M. Sadaoka, H. Azuma, K. Kawamura: Ext. Abstr. 36th Spring Meeting of The Japan Society of Applied Physics, No. 2, p.571 (1989).
12. Rikagaku Jiten (Physics and Chemistry Dictionary) 3rd Edition, Iwanami Shoten (1981).
13. Y. Kawamoto: Data Book on Submicron Lithography, Science Forum, p.335 (1985).
14. Kagaku Binran (Handbook of Chemistry): Maruzen Publishing Co., Ltd.
15. Handbook of Chemistry and Physics 47th Edition: The Chemical Rubber Co. (1966).
16. M. Nakamura, K. Iizuka and H. Yano: Jpn. J. Appl. Phys. 28, 2142 (1989).

Chapter 3
Dry Etching of Various Materials

This chapter reviews the etching of materials actually used in the semiconductor manufacturing process. Etching used in the semiconductor process may be categorized into (1) etching of Si, (2) etching of insulators, and (3) etching of metal line materials. In this chapter, the key technologies in each category—which are gate etching, SiO_2 etching for holes, spacer etching, and etching of Al alloy stacked metal layer structures—are described in detail. The discussions delve beyond simple descriptions and include the information that will help the reader understand the key parameters for the etching process and how to control them. The chapter has been designed so that once the reader understands these etching processes, the knowledge may be applied to the etching of other materials. For example, a section on gate etching reviews the etching of the poly–Si gate, WSi_2/poly–Si gate, and W/WN/poly–Si gate. Once the reader understands these completely, then a similar approach may be followed for designing an etching process for shallow trench isolation (STI) and W metal lines. Also, with gate etching, there is a strong need not only to control the etch profile, but also to minimize the pattern size nonuniformity across the wafer. A discussion on this issue includes which parameters dominate the pattern size uniformity across the wafer and how to control them.

The etching mechanism for SiO_2 material is different from that for the Si material, requiring a different plasma source. This chapter provides an in-depth review of the etching mechanism and the key parameters that dominate the etching process and offers insight into how to design the gas chemistry and the reasons why a narrow-gap parallel-plate etcher is used.

Discussions on Al metal line etching include Al corrosion, which tends to be a problem in the manufacturing process, and how to address it. Incidentally, the Cu damascene technology has become the mainstream metallization alternative to Al interconnect. This topic is covered in Chap. 6.

K. Nojiri, *Dry Etching Technology for Semiconductors*,
DOI 10.1007/978-3-319-10295-5_3

3.1 Gate Etching

Let's first discuss gate etching. The process flow for gate etching is as already described in Chap. 1, Fig. 1.5. A gate is a critical part that determines the transistor characteristics, and the MOS transistor threshold voltage (V_{th}) is especially dependent on gate dimensions. It is vitally important to keep the etch finish dimensions (CD) in control. The physical gate length in the logic devices is 45 nm or less on a 65-nm node and beyond and is approaching 25 nm. Therefore, there is an enormous need not only to control the precision of the CD itself, but also to reduce the CD nonuniformity across the wafer. For example, a gate with a 30-nm CD would require a CD uniformity (3σ) of 3 nm or less across a 300-mm wafer. The etch profile, of course, needs to be vertical, and high selectivity is required because the gate oxide film continues to become thinner with scaling. The gate materials being used are poly–Si for logic devices and multilayer structures such as WSi_2/poly–Si and W/WN/poly–Si for DRAM.

3.1.1 Poly–Si Gate Etching

Conventionally, chlorofluorocarbon or freon-based gases have been widely used for gate etching [1]. However, as we discuss later, etching with chlorofluorocarbon makes it difficult to achieve a high selectivity to the gate oxide film, because carbon in the gas promotes the etching of SiO_2. Furthermore, because of environmental concerns, some types of freon are no longer available for use. Because of these issues, Cl- and Br-based gases are more widely used today. Specifically, the gases used are based on Cl_2 or HBr. As discussed in Sect. 2.3.2 of Chap. 2, etching with the Cl and Br gases tends to yield anisotropic profiles and is suited for achieving a vertical etch profile. Furthermore, as we describe later, a high selectivity to the underlying gate oxide film can be achieved.

Table 3.1 shows the bond strengths for Si, SiO, halogenated Si, and CO [2]. The pre-etch bond strength (Si–O, Si–Si) shown here are for crystals, while the values for post-etch (C–O, Si–F, Si–Cl, Si–Br) are for diatomic molecules. Since the bond strengths of the Si–Cl bond and Si–Br bond are smaller than that of the Si–O bond, the SiO_2 etch rate becomes extremely low with Cl and Br. As a result, high selectivity can be achieved. However, when the carbon exists inside the chamber, breaking or

Table 3.1 Bond strength for Si, SiO, halogenated Si, and CO [2]

C–O			(kcal/mol)
	257[a]	Si–F	132[a]
–Si–O–	111[b]	Si–Cl	96[a]
–Si–Si–	54[b]	Si–Br	88[a]

[a]Crystal
[b]Diatomic molecule

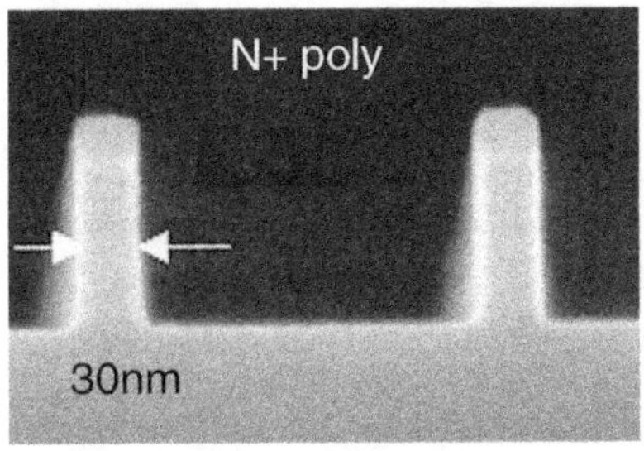

Fig. 3.1 Etching profile of poly–Si gate [3]

weakening of Si–O bonds occurs through the formation of C–O bonds, because the strength of the C–O bond is larger than that of the Si–O bond. As a result, SiO_2 etching would proceed as Si–Cl or Si–Br bonds are created [2]. In this case, the selectivity would drop even with Cl or Br. This is the reason why Cl_2 and HBr tend to give a higher selectivity, while chlorofluorocarbon gives a lower selectivity. Another effective way of achieving anisotropic profiles and a high selectivity ratio is to add O_2 to Cl_2 and HBr. This is discussed in Sect. 3.1.3 on the etching of WSi_2/poly–Si.

Figure 3.1 shows an example of poly–Si gate etching [3]. Cl_2- and HBr-based gas chemistries are used for the etching on a transformer-coupled plasma (TCP®) etcher (this equipment is discussed in Chap. 4). Fine geometry poly–Si gates, with a vertical profile and 30-nm CD, are obtained.

3.1.2 Across-Wafer Critical Dimension Uniformity Control

The across-wafer CD uniformity is strongly affected by the redeposition of reaction byproducts onto the patterns [4]. Figure 3.2 illustrates how the redeposition of reaction byproducts on the patterns affects the CD. Two key parameters determine the CD uniformity. The first parameter is the spatial distribution of the etch byproducts. As shown in the figure, when the concentration of the reaction byproduct $SiCl_4$ (which is generated as poly–Si etched in a Cl_2-based gas) is high at the center of the wafer and low at the peripheries, a large amount of $SiCl_4$ is deposited on the patterns at the center of the wafer, with less at the periphery. As a result, the CD would be larger at the center of the wafer and smaller at the wafer periphery. The other key parameter is the temperature distribution across the wafer. As the figure shows, when the wafer temperature is low at the center and high at the periphery, the sticking probability of $SiCl_4$ is higher at the center of the wafer and lower at the periphery. As a result, the CD becomes larger at the center of the wafer and smaller at the wafer periphery.

Figure 3.3 shows the result of a study on how the direction of gas injection affects the CD uniformity [5]. An etcher used in this study has a function of changing the direction of gas injection. When the gas is injected toward the center of the wafer, the $SiCl_4$ concentration is low at the center of the wafer and high at the periphery, as the figure shows. As a result, the patterns at the center of the wafer become thinner,

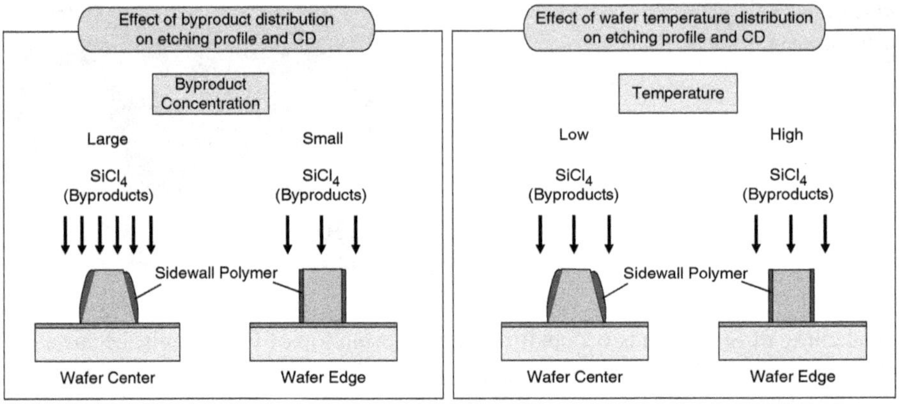

Fig. 3.2 Effect of byproduct and wafer temperature distributions on etching profile and CD

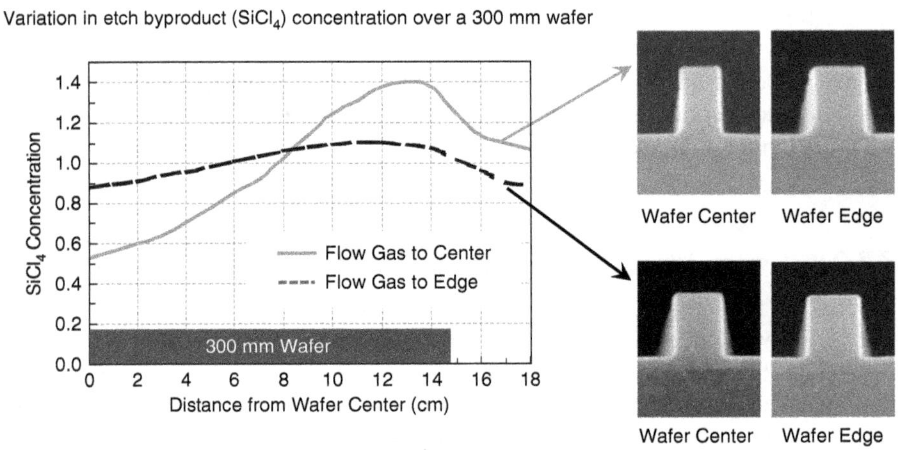

Fig. 3.3 Impact of the direction of gas injection on CD [5]

while the patterns at the periphery are thicker. When the gas is injected toward the periphery of the wafer, the uniformity of the $SiCl_4$ concentration across the wafer improves, and the difference in the CDs becomes smaller between the center and the periphery of the wafer.

Figure 3.4 shows the result of a study on how the temperature distribution across the wafer affects the CD uniformity in an etcher having an electrostatic chuck that is capable of independent temperature control at the center and periphery of a wafer [5]. When the electrostatic chuck temperature is set at 30 °C at both the center and periphery of the wafer, the patterns at the center end up being slightly thicker. When the temperature at the center of the wafer is raised to 50 °C, the patterns at the center end up being thinner. The electrostatic chuck is discussed in Chap. 4, Sect. 4.8.

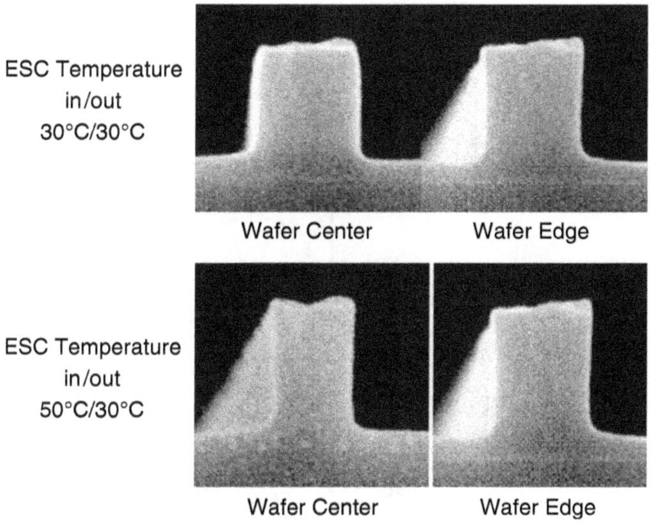

Fig. 3.4 Impact of electrostatic chuck (ESC) temperature on CD [5]

As discussed, it is necessary to precisely control the redeposition of byproducts onto the patterns in order to suppress the CD variation across the wafer and achieve a better uniformity. In order to achieve this goal, it is necessary to use (1) an etcher having an injector that enables gas injection in a specific direction and (2) an electrostatic chuck that is capable of adjusting the temperature distribution across the wafer. By optimizing the two key parameters described earlier, a 3σ uniformity of 2.5 nm has been achieved across a 300-mm wafer [6].

3.1.3 WSi_2/Poly–Si Gate Etching

A WSi_2/poly–Si structure is referred to as a polycide gate and is widely used in DRAMs. A DRAM gate is a word line, and WSi_2 is used here to reduce the word line resistance. With the WSi_2/poly–Si structure, it is necessary to etch vertically without any side etching into the WSi_2 or poly–Si. So the process is very challenging. This section describes an example of an etching process with O_2 added to Cl_2 [7]. The etching equipment used here is an electron-cyclotron resonance (ECR) plasma etcher, and the etching mask is a chemical vapor deposition (CVD) SiO_2.

Figure 3.5 shows the O_2 concentration dependence of the etch rate and selectivity [7]. The etch rate of SiO_2 decreases as the O_2 concentration increases, and the poly–Si/SiO_2 selectivity goes up. Furthermore, the etch rate of WSi_2 increases as the O_2 concentration increases up to 10 % and has the same value as that of poly–Si. The WSi_2 etch rate increases by adding O_2 because the vapor pressure of $WOCl_4$ is higher than that of the WCl_6 [8]. The selectivity is further increased when the ion energy is reduced by reducing the RF power.

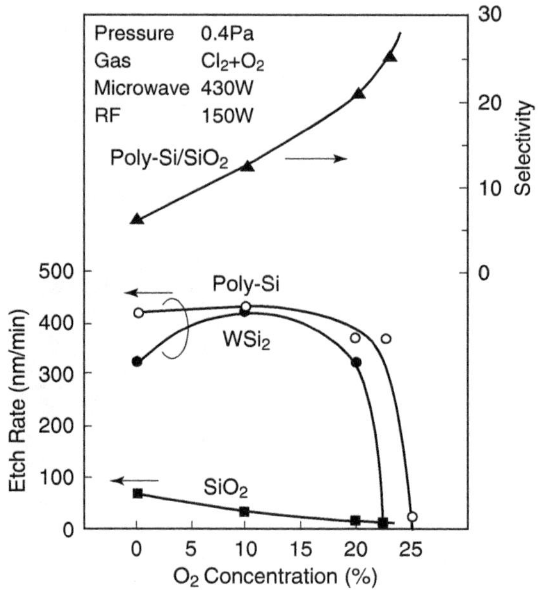

Fig. 3.5 Etch rate and selectivity as a function of O_2 concentration [7]

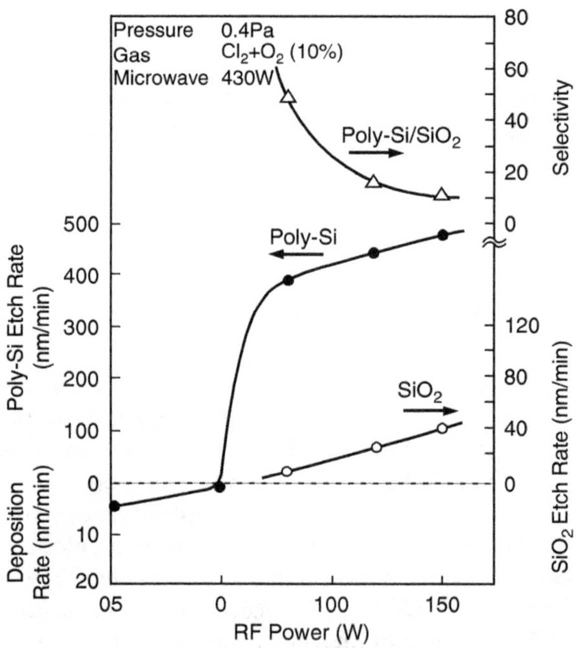

Fig. 3.6 Etch rate and selectivity as a function of power [7]

Figure 3.6 shows the RF power dependence of the etch rate and selectivity, at an O_2 concentration of 10 % [7]. A negative etch rate indicates the deposition of a film. The etch rate of SiO_2 drops drastically at a lower RF power. On the other hand, the poly–Si etch rate does not show a significant drop down to an RF power of 80 W. As a result, the poly–Si/SiO_2 selectivity drastically increases when the RF power

RF Power	80 W	80 W
Etching Gas	Cl_2	Cl_2+O_2 (10%)
形状		
Etch Rate	350 nm/min	400 nm/min
Selectivity (Poly-Si/SiO_2)	9	50

Fig. 3.7 Scanning electron microscope cross-sectional profile of WSi_2/poly–Si gate etched with Cl_2 and Cl_2/O_2 (10 %) [7]

is decreased. The poly–Si/SiO_2 selectivity and the poly–Si etch rate are 50 and 400 nm/min, respectively, at an RF power of 80 W. The etching of poly–Si does not take place below 50 W, and a thin film is deposited on the wafer surface. The deposition rate of the thin film is 4 nm/min at an RF power of 0 W. This result indicates that the sidewall protection film is deposited on the sidewall of WSi_2/poly–Si etched with $Cl_2 + O_2$, because the etch rate at the RF power of 0 W is approximately equivalent to the etch rate at the sidewall where few ions impinge.

Figure 3.7 compares the etching profiles resulting from Cl_2 alone, or with an O_2 concentration of 0 %, and with 10 % O_2 added to Cl_2 [7]. The RF power is 80 W. While side etching is observed with an O_2 concentration of 0 %, a vertical etching profile with no side etching is achieved when 10 % O_2 is added because the byproducts, resulting from the O_2 addition, would deposit on the sidewalls and protect them. The composition of the sidewall protection film can be estimated by analyzing the film deposited on the poly–Si surface at an RF power of 0 W in Fig. 3.6, because, as mentioned earlier, the phenomenon at the RF power of 0 W approximately represents the phenomenon that takes place on the sidewalls. Figure 3.8 shows the result of an Auger electron spectroscopy analysis on this film. Strong peaks for O and Si are detected, indicating that the film consists of O and Si. This result shows that the sidewall protection film is SiO_x. As indicated in Table 3.1, since the bond strength of the Si–O bond is larger than that of the Si–Cl bond, Cl radicals barely etch the sidewall protection film. Therefore, the sidewall protection film effectively protects the sidewalls from the attacks by Cl radicals. As a result, an anisotropic etching is achieved.

The reduction in the SiO_2 etch rate, when O_2 is added, is also thought to be due to the deposition of SiO_x. According to Fig. 3.5, etching stops when the O_2 concentration exceeds 25 % and a deposited film is observed on the wafer surface. An Auger electron spectroscopy analysis shows that this deposited film is also SiO_x [7].

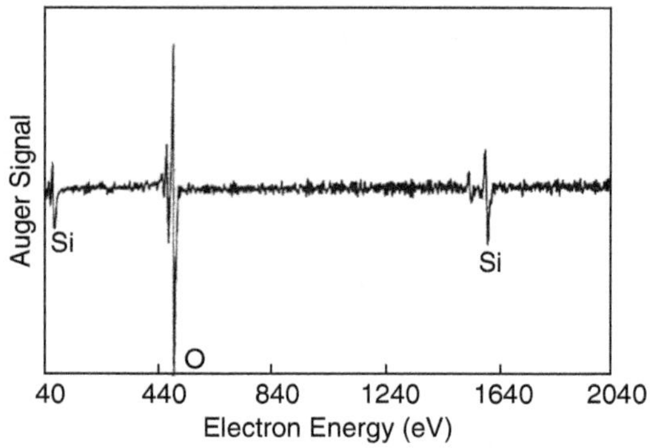

Fig. 3.8 Auger electron spectroscopy analysis of the film deposited on the poly–Si surface at an RF power of 0 W and an O_2 concentration of 10 % [7]

These results would lead to a theory that the reduction in the SiO_2 etch rate, due to the O_2 addition, is also caused by the SiO_x deposition.

As discussed above, the O_2 addition is effective not only for increasing the WSi_2 etch rate and the poly–Si/SiO_2 selectivity, but also for forming a sidewall protection film to achieve an anisotropic etching.

3.1.4 W/WN/Poly–Si Gate Etching

With the scaling of DRAM, WSi_2 has been replaced by W, which offers a lower resistivity. The gate structure used today is W/WN/poly–Si, which is referred to as a poly–metal gate. Etching on this structure is possible through a similar approach taken for the WSi_2/poly–Si, as discussed earlier. But etching with Cl_2 is more difficult in this case compared with etching on WSi_2, because Si is not contained in W. An effective approach here would be basically either to increase the vapor pressure of the byproduct (tungsten chloride) by increasing the wafer temperature [9] or to add a fluorine-based gas in order to generate the tungsten fluoride, which has higher vapor pressures. Figure 3.9 shows an example of high-temperature etching on an ECR plasma etcher [9]. The W/WN was etched at 100 °C, with $Cl_2 + O_2$ (12 %) gas chemistry. A vertical profile with no side etching is achieved.

3.1.5 Silicon Substrate Etching

Representative examples of etch processes on the Si substrate are STI and through-Si via (TSV). As discussed in the DRAM manufacturing process flow in Chap. 1, STI is used for isolation between the devices. It is used in both logic devices and memory devices. The etching method used for poly–Si can basically be applied directly to the STI. Cl_2- or HBr-based etching chemistry is used. A TSV is the through-hole via used for creating 3D devices. It is an etching process to form high-aspect-ratio deep holes into Si. This is discussed later in Chap. 6.

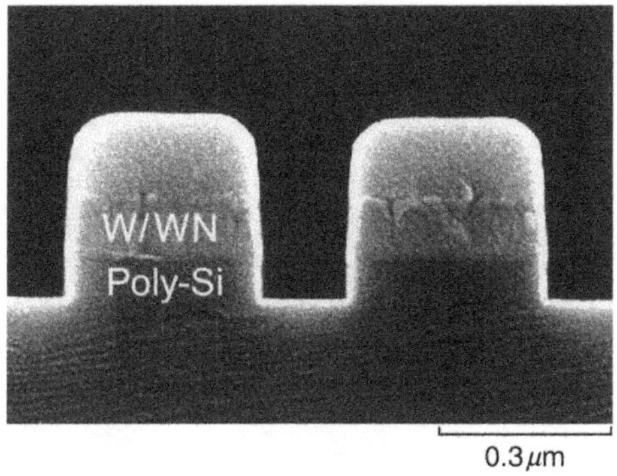

Fig. 3.9 High-temperature etching of W/WN/poly–Si gate [9]. (W/WN was etched at 100 °C)

3.2 Silicon Dioxide Etching

There is a wide variety of applications in SiO_2 etching, including the etching of holes such as contact holes and via holes, hard mask etching for the STI and gates, and damascene etching, which is discussed in Chap. 6. Also, the number of process steps in device manufacturing is the largest among the etching technologies.

Compared with conductor material etching, such as Si, poly–Si, and Al etching, SiO_2 etching has a complicated etching mechanism and requires a different type of plasma source. Let's first review the etching mechanism.

3.2.1 Mechanism of Silicon Dioxide Etching

For SiO_2 etching, C and F are necessary, and the etching gas is based on fluorocarbon containing C and F. Furthermore, H_2- or H-contained gas is added to the fluorocarbon gas in order to obtain a large selectivity to the underlying Si. The reason for this mixture is described next.

Figure 3.10 illustrates the etching mechanism when SiO_2 is etched with a CF_4/H_2 mixture gas [10]. The CF_4 dissociates into CF_3^+ ions, CF_3 radicals ($CF_3\bullet$), and F radicals ($F\bullet$) in the plasma. The H_2 generates H radicals ($H\bullet$). Let's first examine the reactions at the SiO_2 surface. The CF_3 radicals adsorbed on the SiO_2 surface are dissociated into C and F under the irradiation of CF_3^+ ions. As discussed in Sect. 3.1.1, since the C–O bond strength is larger than the Si–O bond strength, C reacts with O in SiO_2 to make CO, and then CO desorbs from SiO_2 surface. The Si, which in turn is weakly bonded, reacts with F to form SiF_4, and then SiF_4 desorbs from SiO_2 surface. Etching of SiO_2 thus proceeds. In other words, because SiO_2 contains O atoms, C, which reacts with O and forms volatile material, must always be included in the etching gas. These are the reasons why a fluorocarbon containing F and C is used as the base gas.

Let's now look at the reactions taking place at the Si surface. Hydrogen radicals extract F from CF_3 adsorbed on the Si surface, and as a result, fluorocarbon

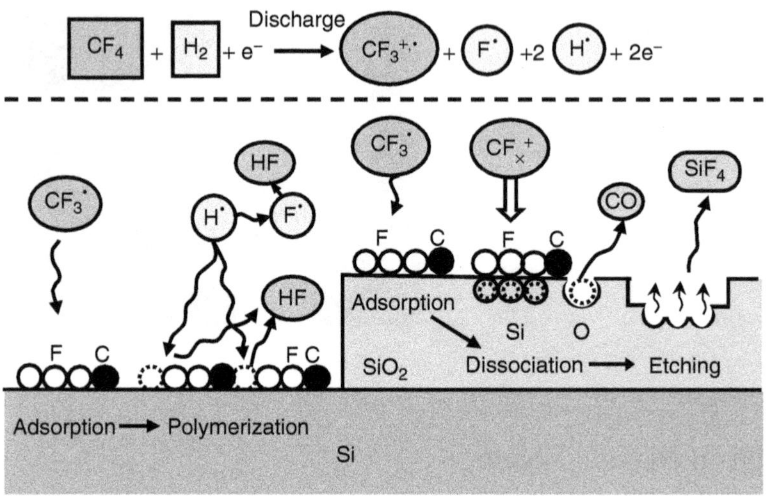

Fig. 3.10 Model for SiO_2 etching with CF_4+H_2 gas chemistry [10]

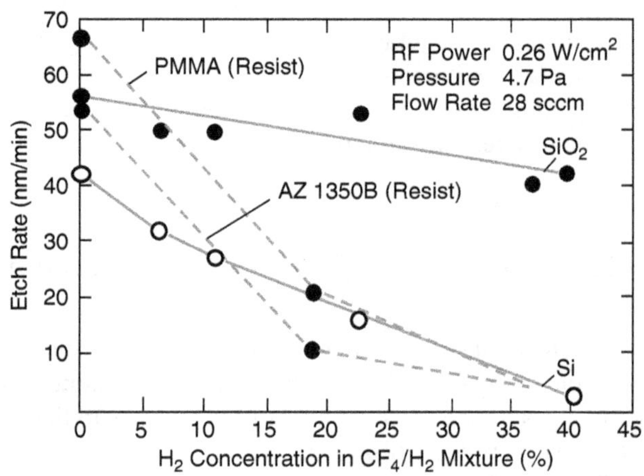

Fig. 3.11 SiO_2, Si, and resist etch rates as a function of H_2 concentration [11]

polymers are formed. Because the Si surface is covered with polymer, the etching of Si by F radicals is suppressed. Furthermore, the H radicals react with the F radicals in the gas phase to form HF and reduce the number of F radicals, which are the etchant of Si. This is referred to as the "scavenging effect." Due to these two effects, the etch rate of underlying Si drops, and etching with a high SiO_2/Si selectivity is realized. These are the reasons why H is required in the etching gas. Figure 3.11 shows the SiO_2, Si, and resist etch rates as a function of the H_2 concentration in the CF_4/H_2 mixture [11]. The Si etch rate drastically drops as the H_2 concentration increases. Because the SiO_2 etch rate decreases only slightly, the SiO_2/Si selectivity increases with an increased H_2 concentration. The SiO_2/Si selectivity improves to approximately 10 at an H_2 concentration of 40 %. The resist etch rate shows a similar trend, and the SiO_2/resist selectivity is also around 10 at an H_2 concentration of 40 %.

Similar effects to those described here may also be achieved using gases such as CHF_3, which itself contains H. Trifluoromethane (CHF_3) is widely used as an SiO_2 etching gas. Furthermore, because the selectivity tends to increase in gases with a larger C ratio, as described in the next section, fluorocarbons such as C_2F_6, C_3F_8, C_4F_8, and C_5F_8 have also been used in recent years.

3.2.2 *Key Parameters in Silicon Dioxide Etching*

The key parameters in the SiO_2 etching are discussed next. An important parameter that determines the SiO_2/Si selectivity and SiO_2/resist selectivity is the CF_2/F ratio. Selectivity becomes larger with a larger CF_2/F ratio [12]. Figure 3.12 shows the results of an experiment conducted to study what determines the F/CF_x ratio [13]. In this experiment, $C_4F_8/O_2/Ar$ gas chemistry was used. As the figure shows, the F/CF_x ratio depends on $\tau \cdot n_e$, which is a product of the residence time of the particles τ and the electron density (plasma density) n_e, and increases in proportion to $\tau \cdot n_e$.

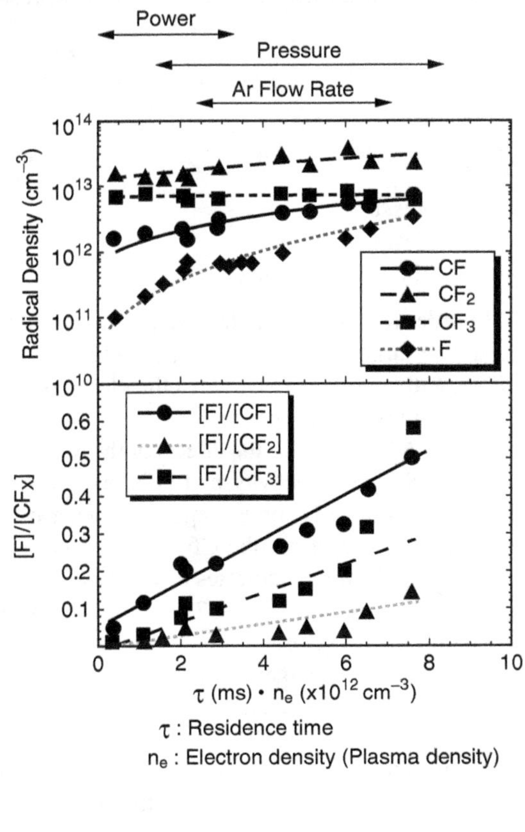

Fig. 3.12 The CF, CF_2, CF_3, and F radical densities, and F/CF_x ratio in $C_4F_8/O_2/Ar$ plasma as a function of $\tau \cdot n_e$ [13]

The number of collisions between the electrons and molecules can be expressed by the following equation:

$$\xi = \tau \cdot n_e < \sigma v >, \tag{3.1}$$

where

τ is the residence time of the particles,
n_e is the electron density (plasma density),
σ is the collision cross section,
ν is the electron velocity.

The number of collisions increases as the residence time and/or electron density (plasma density) increases. As a result, dissociation of C_4F_8 proceeds to create more F, and the F/CF_x ratio becomes larger. The CF_2/F ratio decreases as τ and/or n_e increases. This leads to a lower selectivity for SiO_2/Si and SiO_2/resist. In other words, in SiO_2 etching, a higher selectivity can be achieved through a shorter residence time and a lower plasma density. Therefore, a high-density plasma is not suitable for SiO_2 etching, and a medium-density plasma is generally used. As is described in further detail in Chap. 4, a narrow-gap parallel-plate etcher with medium-density plasma is the suitable etching equipment.

The residence time of the particle τ is expressed with the following equation:

$$\tau = \frac{PV}{Q}, \tag{3.2}$$

where

P is the pressure,
V is the plasma volume,
and Q is the gas flow rate.

In order to make τ smaller, the plasma volume should be small, the gas flow rate should be large, and the pressure should be low. Using a narrow-gap parallel-plate etcher, with a small gap between the electrodes in the etch chamber, would result in a smaller plasma volume.

Figure 3.13 illustrates the effects of the distance between the electrodes on C_4F_8 dissociation, SiO_2/Si selectivity, and SiO_2/resist selectivity. As the electrode gap increases, the plasma volume becomes large, and then τ becomes large. This leads to more C_4F_8 dissociation. As a result, the CF_2/F ratio decreases, and the selectivity goes down. In other words, the figure indicates that a narrow electrode gap would be better for achieving a higher selectivity. Figure 3.14 illustrates the effect of Ar flow rate on the C_4F_8 dissociation, SiO_2/Si selectivity, and SiO_2/resist selectivity. When the pressure is constant, τ decreases with the larger Ar flow rate, according to Eq. (3.2), and the C_4F_8 dissociation is suppressed. As a result, the CF_2/F ratio increases, and selectivity goes up.

The double-near-surface model [14] shown in Fig. 3.15 provides another explanation for the reason why a narrow-gap parallel-plate etcher is suitable for the SiO_2 etching. According to this model, the reactions taking place at the upper electrode

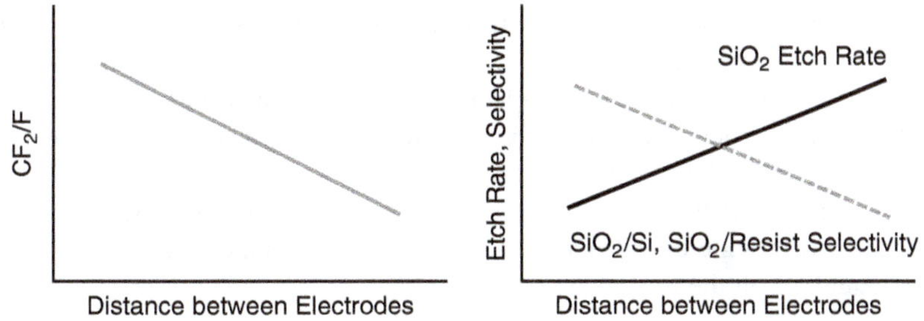

Fig. 3.13 Impact of distance between electrodes on the dissociation of C_4F_8, and SiO_2/resist selectivity

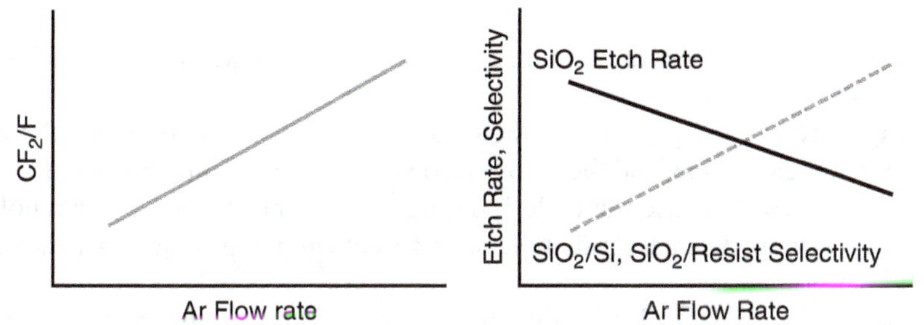

Fig. 3.14 Impact of Ar flow rate on C_4F_8 dissociation, and SiO_2/Si and SiO_2/resist selectivity

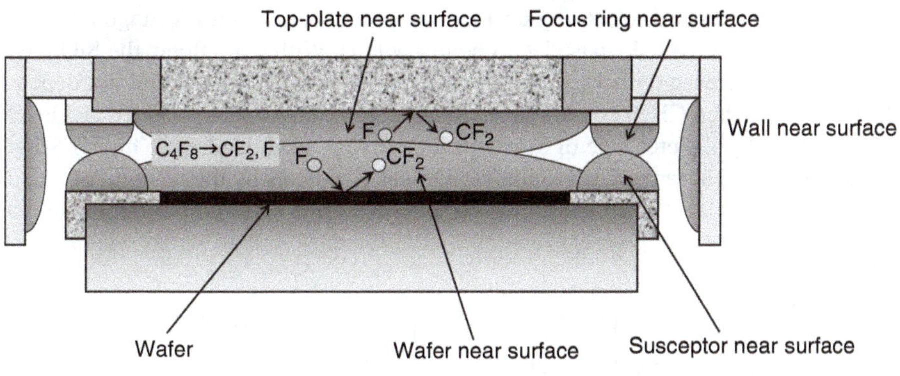

Fig. 3.15 Double-near-surface model for narrow-gap–type SiO_2 etcher [14]

consume F and release CF_2. Furthermore, the same reactions are taking place at the wafer surface. In other words, the CF_2 concentration is higher near the upper electrode and the wafer. Therefore, when the electrode gap is narrower, these regions would overlap, and the CF_2 concentration would further increase. As a result, the selectivity goes up. This is another reason why the narrow-gap parallel-plate etcher is suitable for SiO_2 etching. The electrode gap is usually set at 20–30 mm.

It has been reported that the reaction resulting in the release of CF_2 is accelerated with temperature and biasing [15, 16]. The SiO_2/Si selectivity and the SiO_2/resist selectivity improve when the upper electrode is heated or a bias is applied on the upper electrode. At one time, there was an effort to use high-density plasmas, such as inductively coupled plasma (ICP) and ECR, for the SiO_2 etching. However, they could not achieve the high selectivity, and in the end the narrow-gap parallel-plate etchers came into wide use. The descriptions earlier in this chapter should provide an understanding on the reason for this outcome.

3.2.3 *Hole Etching*

The next discussion covers the etching of holes, such as contact holes, via holes, and DRAM capacitor cells.

As Fig. 3.16 shows, gate contact holes are shallow and substrate contact holes are deep in a logic device, and the simultaneous etching of contact holes with varied depths is required. Because the underlying material in the shallow contact holes is exposed to a long plasma irradiation, a large selectivity is required against the underlying poly–Si.

As described in Sect. 3.2.1, high SiO_2/Si selectivity is obtained by forming polymers on the Si surface, because they decrease the Si etch rate. In order to form the polymers on the Si surface, it is important to control the ion acceleration voltage in addition to the concentration of the additive gases. Figure 3.17 shows the dependence of etch rates and selectivity on V_{max} (maximum acceleration voltage for ions) for the etching with a C_4F_8-based gas chemistry [17]. With C_4F_8 alone, the SiO_2 and Si etch rates increase with V_{max}, but the SiO_2/Si selectivity is small and not dependent on V_{max}. When CH_3F containing a large amount of hydrogen is added at a level of 30 % to C_4F_8, the etch rate of Si is considerably reduced, leading to higher SiO_2/Si selectivity, because the CH_3F addition reduces F radicals by the reaction of F and H to form HF as well as enhances the polymerization of CH and CF fragments on

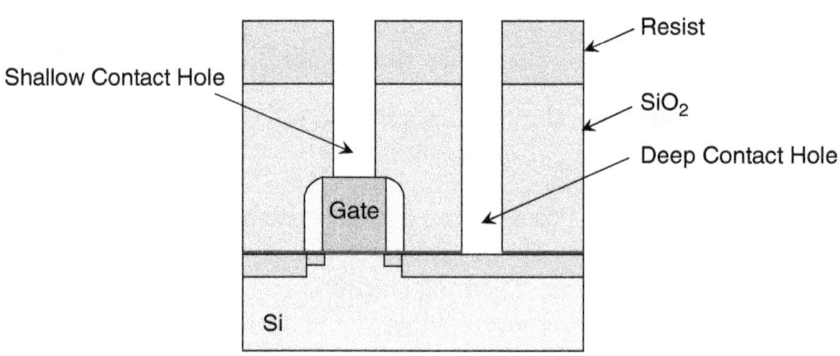

Fig. 3.16 Contact hole etching for logic device

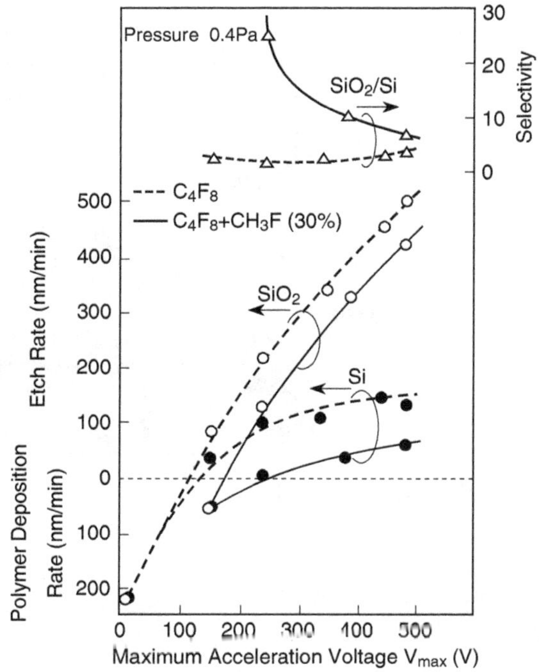

Fig. 3.17 The SiO_2 and Si etch rate and Si SiO_2/Si selectivity as a function of maximum acceleration voltage V_{max} [17]

the Si surface. This effect becomes more significant with a lower V_{max}. When V_{max} is 200 V or less, the Si surface is completely covered with polymer, and Si etching stops; that is, the selectivity is infinite. However, because the SiO_2 etch rate also drops, it is necessary to select the optimum ion energy in an actual process, by taking into account the tradeoff between the etch rate and selectivity.

With DRAM, the aspect ratio (depth/hole diameter) of the contact holes and capacitor cells is extremely large, as Fig. 3.18 shows. The aspect ratio is even larger when the resist mask is included. Contact holes with a high aspect ratio are referred to as high-aspect-ratio contacts (HARC). The aspect ratio of HARC for the 50-nm-node DRAM, currently in volume production, is around 17.

High-aspect-ratio contact etching first requires a low-pressure regime. As previously discussed in Sect. 2.2.2 in Chap. 2, this is required for making the ions travel in a straight direction and enter into small holes. Bowing can be a problem with HARC etching. It is a phenomenon in which the etch profile becomes barrel-shaped because of the side etching at the middle of the hole. Figure 3.19 illustrates the mechanism [18]. Bowing is caused by the ions scattered at mask edges and hitting the sidewalls of the holes. Deposition of the polymers on the sidewalls is driven by CF_2, which has a small sticking coefficient of 0.004 and tends to reach the bottom after reflecting multiple times on the sidewalls. C, which has a large sticking coefficient of 0.5, turns into polymers on the resist mask and at the upper part of the hole [18]. As shown in the figure, the polymer depositing at the upper part of the hole increases the rate of ion scattering. Therefore, bowing can be reduced by

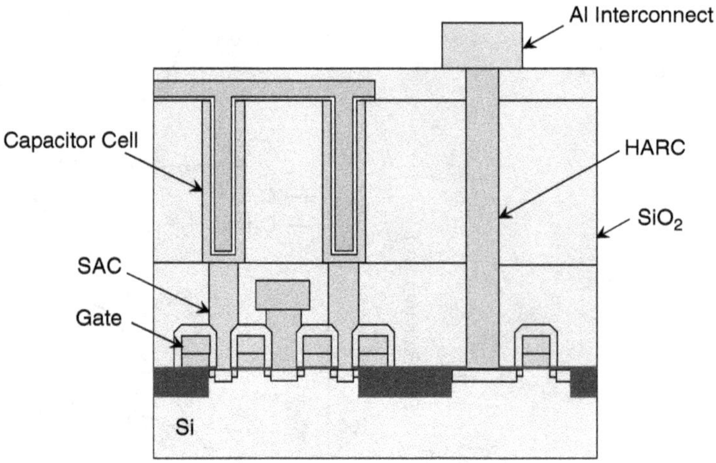

Fig. 3.18 Cross-sectional view of DRAM

Fig. 3.19 Mechanism of bowing formation in HARC etching [18]

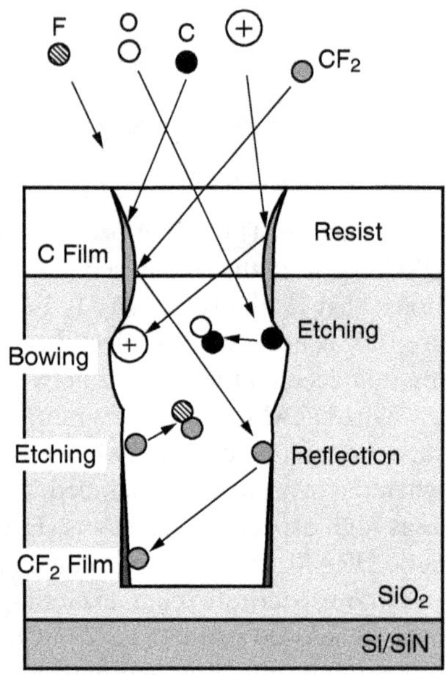

the following measures [18]. First, deposit thicker CF_2 polymer on the sidewall by increasing the CF_2 concentration in the plasma. The same approach described earlier should be utilized. Next, transport more C to the bottom by reducing the sticking coefficient with a higher wafer temperature. Figure 3.20 shows a cross-sectional SEM photograph of the HARC etched with $C_4F_8/O_2/Ar$ gas chemistry on a narrow-gap parallel-plate etcher [13]. No bowing is observed for the hole with a 0.1-μm diameter and an aspect ratio of 10. It has also been reported that holes with a 0.08-μm diameter and an aspect ratio of 19 were successfully etched without bowing when the same etching gas chemistry was used [19].

Fig. 3.20 Cross-sectional SEM photograph of the HARC etched with $C_4F_8/O_2/$ Ar gas chemistry [13]

3.2.4 *Self-aligned Contact Etching*

Self-aligned contact (SAC) etching is a technology for opening contact holes between the gate electrodes, as Fig. 3.21 shows. The gates are covered with an Si_3N_4 film, which functions as an etch stopper, so that the contact hole and the gates will not be shorted even when there is a misalignment. It is then possible to expand the alignment margin and scale down the chip size. This is an essential technology for DRAM. The key challenge in SAC etching is the maximization of SiO_2/Si_3N_4 selectivity. A gas mixture of C_4F_8 and CO is frequently used for SAC etching. Figure 3.22 shows the measurement results on etching yields when the SiO_2 and Si_3N_4 are irradiated with CF_2^+ and $C_2F_4^+$ ions [20]. Here the etching yield is defined as the quotient of the number of etched Si atoms divided by the number of incoming ions. The SiO_2 etching yield is much higher for $C_2F_4^+$ compared to CF_2^+, and the SiO_2/Si_3N_4 selectivity is higher. When the ion energy is reduced to 200 eV, the polymer deposition is enhanced. As a result, the Si_3N_4 etch rate decreases significantly, and an SiO_2/Si_3N_4 selectivity of 80 is obtained. Figure 3.23 shows the measurement results on the ion species concentration in the C_4F_8, C_4F_8/CO, and C_4F_8/H_2 plasmas [20]. When CO is added to C_4F_8, the CF_2^+ concentration drops, and the $C_2F_4^+$ concentration increases. It has also been verified that in etching with a magnetron reactive-ion etching (RIE, the equipment described in Chap. 4), the SiO_2/Si_3N_4 selectivity improves with CO addition to C_4F_8, and a selectivity of 15 has been obtained [20].

Octafluorocyclopentene (C_5F_8) has also gained attention as an alternative gas to C_4F_8. There has been a report of a selectivity of 21 at the Si_3N_4 shoulder using a C_5F_8/O_2 gas chemistry in the magnetron RIE [21]. A high selectivity is believed to be achieved through the deposition of C-rich polymers with three-dimensional C–C bonds that have a high plasma resistance.

Fig. 3.21 Self-aligned contact (SAC) etching

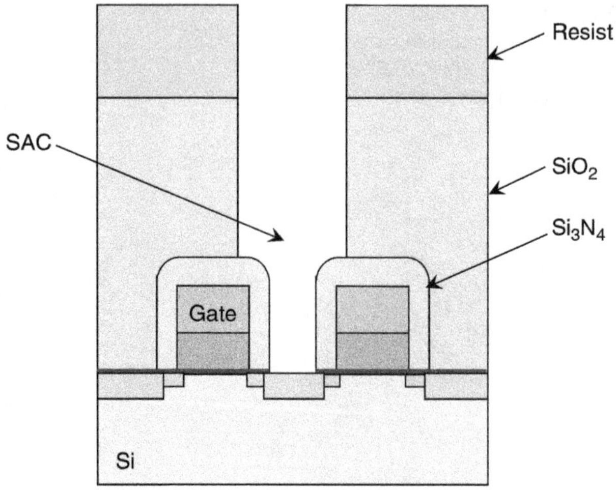

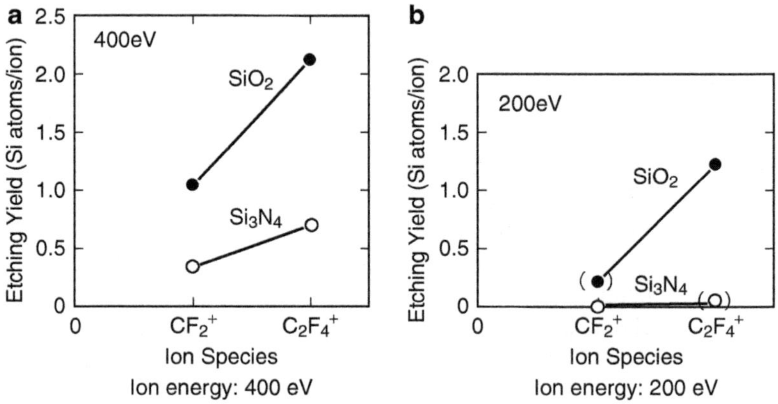

Fig. 3.22 Etching yield of SiO_2 and Si_3N_4 irradiated with CF_2^+ and $C_2F_4^+$ ions [20]

3.2.5 Spacer Etching

The final discussion here is on spacer etching. Spacer etching is an etching process for forming insulating film spacers on the gate sidewalls. As mentioned in Chap. 1 during the discussion on the DRAM process flow, spacers are used for forming the source and drain regions at locations slightly away from the gate. Spacers are also used for double patterning, to be discussed in Sect. 6.6 in Chap. 6. Spacer etching is a process that utilizes the characteristics of anisotropic etching. A process flow is shown in Fig. 3.24 and described as follows:

1. A gate is formed. 2. Then a dielectric film, such as SiO_2 or Si_3N_4, is deposited by CVD. 3. When an anisotropic etching then follows, the CVD insulating film remains on the gate sidewalls without being etched, because the thickness of the CVD dielectric film is larger in the vertical direction, and spacers are thus formed.

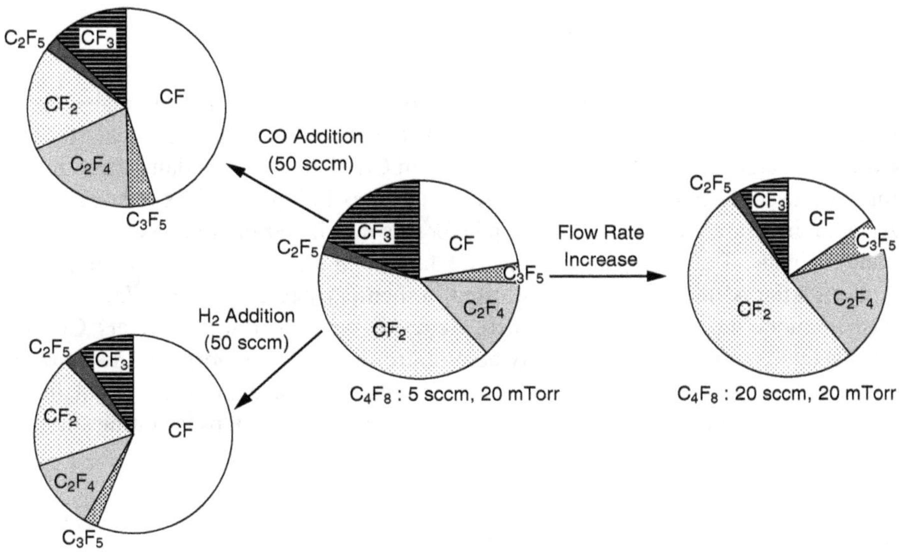

Fig. 3.23 Relative concentration of chemical species in C_4F_8, C_4F_8/CO, and C_4F_8/H_2 plasma [20]

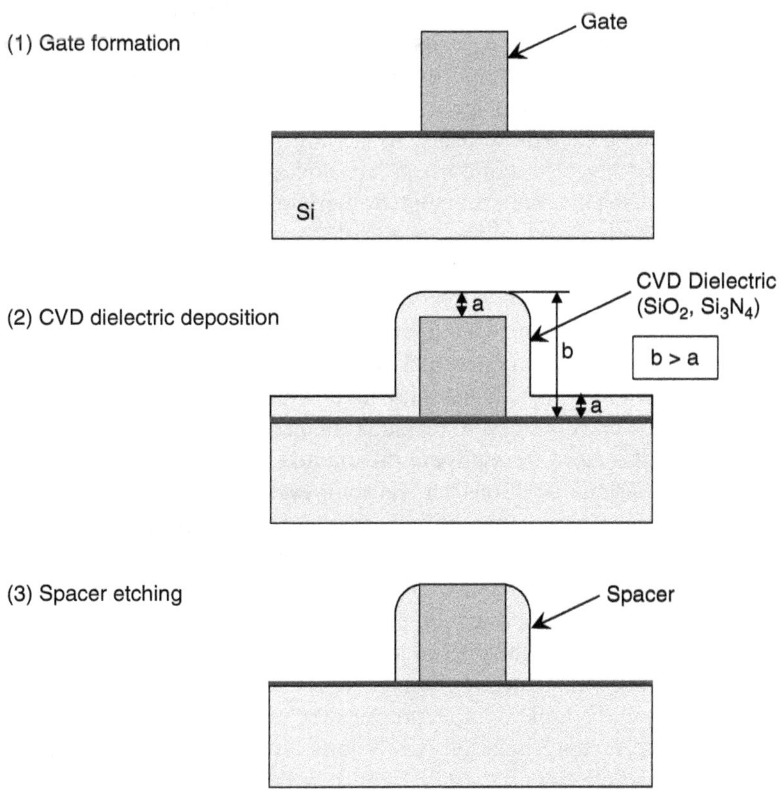

Fig. 3.24 Process sequence of spacer etching

3.3 Metal Line Etching

Multilayer interconnects used in logic devices sometimes contain as many as 10 layers of metal. For this reason, the ratio of metal steps to entire process steps is extremely large. As the detailed discussions in Chap. 6 show, Cu damascene metal lines are used as an alternative to the Al metal lines in logic devices today, and the Al metal lines are only used for the top layer of a multilayer interconnect and for the bonding pads. Because pattern sizes are larger, fine processing is not required. Although the Al metal lines are still used in memory devices, such as DRAM and flash memory, the number of layers here are fewer, typically only 2–3 layers. Copper damascene metal lines are gradually being introduced for memory devices as well, and at least one layer has already been replaced by Cu damascene metal lines. Therefore, the ratio of Al etching steps to all etching process steps is getting smaller.

3.3.1 Aluminum Metal Line Etching

As previously discussed in Sect. 2.3.4 in Chap. 2, Cl_2-based gases are used for Al etching. Aluminum itself is etched at a very high rate by Cl radicals. Because Al tends to be oxidized easily, however, its surface is normally covered with the oxide film (Al_2O_3), which blocks etching. In order to remove this oxide on the surface, BCl_3, which is a strong reducing agent, is usually mixed in. Therefore, the basic gas system used would be Cl_2+BCl_3 for the Al etching.

A stacked metal layer structure of an Al alloy and a barrier metal is used for electromigration-resistance enhancement. Sometimes a thin film is formed on top of the Al alloy as an antireflection film. Specifically, the structure is a sandwich of, for example, TiN/Al–Si–Cu/TiN or TiW/Al–Si–Cu/TiW.

In Al etching, the phenomenon where the amount of side etching varies depending on the patterns is an issue. This phenomenon is referred to as a microloading effect. Figure 3.25 shows experimental results on the microloading effect [22]. A test pattern shown in Fig. 3.25a has a layout in which a long and narrow line, with a width of 4 μm, extends out of a rectangular-shaped pattern having dimensions of 75 mm × 45 mm. It is used for studying the amount of Al side etching (amount of the undercut) at various positions on the long and narrow line. The results are shown in Fig. 3.25b. The equipment is the parallel-plate RIE, and the pressure is 47 Pa. Figure 3.25b shows that the amount of side etching increases as the distance from the rectangular pattern increases. As mentioned in Sect. 2.3.3 in Chap. 2, it is necessary to use the sidewall protection process for the etching of Al with Cl. Since the supply of polymers from the resist declines with more distance from the rectangular pattern, the sidewall protection film becomes thinner. As a result, side etching occurs. In the figure, the broken line represents the result for a UV-cured resist, and the amount of side etching is larger. Ultraviolet cure refers to the heating of the resist under an irradiation with the UV light. It is believed that UV cure results in an increase in Al/resist selectivity, which then further decreases the supply of

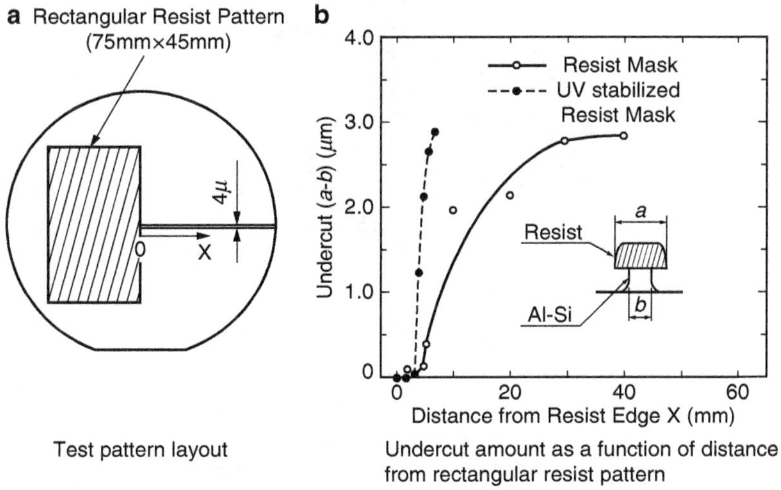

Fig. 3.25 Microloading effect in Al–Si etching [22]

polymers and increases the amount of side etching. This phenomenon, in which the side etching is caused by the decline in the supply of polymers from the resist, is observed on the actual device patterns in which the area of the resist mask is smaller compared to the area that is to be etched. This phenomenon becomes significant in a high-pressure regime. In order to prevent side etching, the operating pressure should be reduced so that more ions travel in a straight direction. In this case, an anisotropic etching is possible even when the sidewall protection films are thinner. The ECR plasma and ICP plasma are effective etching methods, because their operation pressure is low.

Figure 3.26 shows the effect of resist on the etching profile and side etching. Here half of a wafer is coated with the resist, and the profile and side etching are investigated as a function of the distance from this area [23]. An ECR plasma etcher is used, and the pressure is 2 Pa. Both the etching profile and the amount of side etching show almost no changes over a distance ranging from 5–40 mm from the area covered by the resist. Another way of preventing side etching is to enhance the strength of the sidewall protection film with a gas additive. For example, when a CH-based gas is added, polymers, similar to the resist, deposit on the sidewalls and enhance the sidewall protection film.

Figure 3.27 shows the V_{dc} dependence of the Al–Si–Cu etch rate and the side etching amount in an ECR plasma [24]. The pressure is 1.3 Pa. The figure shows that it is possible to control the amount of side etching without changing the etch rate, by adjusting the V_{dc}. Figure 3.28 shows an example of a stacked metal layer structure etched with an ECR plasma etcher [25]. Anisotropic etching with no steps among the various layers is realized. Among the barrier metals, TiN is relatively easy to etch, because it can be etched using Cl_2. On the other hand, TiW requires an F-based gas for etching because 80–90 % of its composition is W, and the etching characteristics of this material are similar to those of W.

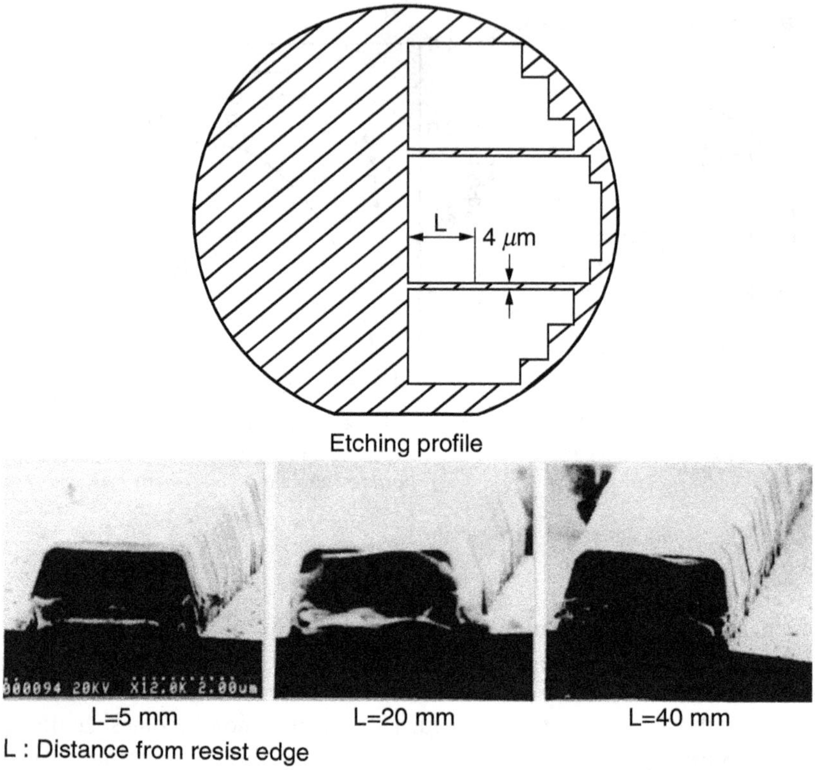

Fig. 3.26 Effect of resist on Al–Si–Cu etching profile [23]

Fig. 3.27 The Al–Si–Cu etch rate and side etching amount as a function of V_{dc} [24]

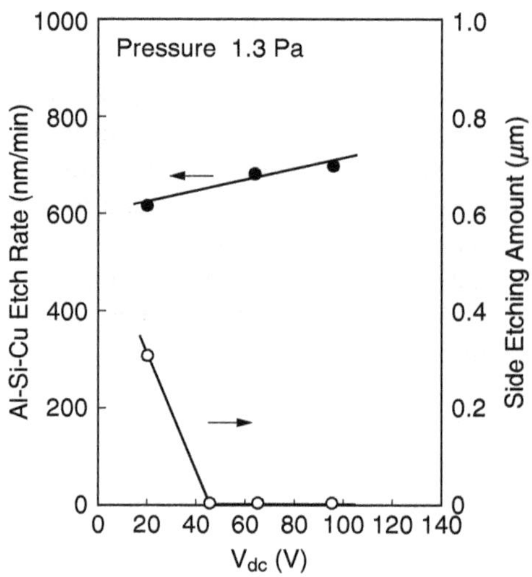

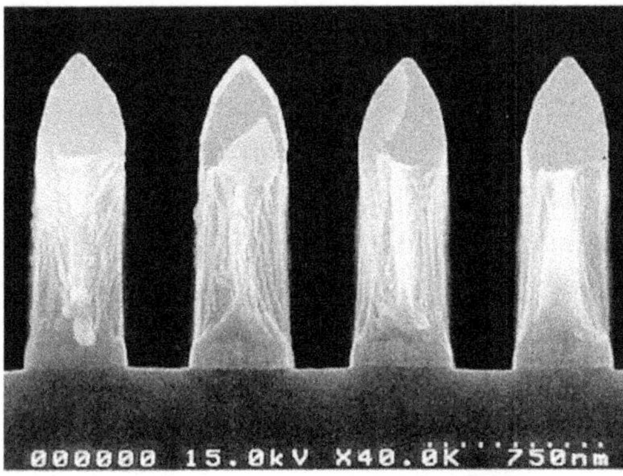

Fig. 3.28 Etching profile of Al–Si–Cu [25]

3.3.2 Anticorrosion Process in Aluminum Etching

Another challenge in Al etching is the anticorrosion process. There would be a significant amount of corrosion if the wafer were taken into the atmosphere with a large amount of residual Cl on it after Al etching. This is because HCl is generated by the reaction of the moisture in the atmosphere and the residual Cl. With an Al alloy containing Cu, the Cl^- ions would easily destroy the Cu oxide film, and corrosion progresses by galvanic effect. The potential for corrosion is further increased with stacked layer structures such as TiN/Al–Si–Cu/TiN and TiW/Al–Si–Cu/TiW because the galvanic effect is enhanced. In any case, it is important to completely remove the residual Cl on the wafer in order to prevent corrosion [26].

Figure 3.29 shows the amount of residual Cl on the wafer as a function of process step of Al–Si–Cu etching [27]. More than 99 % of the residual Cl can be removed when the resist is stripped off by an asher. When this step is followed by a wet cleaning, the Cl concentration goes back to almost the pre-etching level. When the etching system is equipped with an in-line asher that is able to strip off the resist without breaking the vacuum, then it is possible to prevent corrosion only by resist stripping, as long as the Cu content in the Al alloy is around 1 %. An effective approach on a stacked structure with barrier metal would be to add a further cleaning step with deionized (DI) water.

3.3.3 Etching of Other Metal Materials

Other metal materials that need to be mentioned are W and Cu. W is used for parts of memory metal lines. An approach similar to those discussed in Sect. 3.1.4 on the W/WN/poly–Si gate can be used for W etching.

The etching of Cu is extremely difficult. As shown in Table 2.3 in Chap. 2, Cu halides have an extremely low volatility. Therefore, in order to etch the Cu, it is

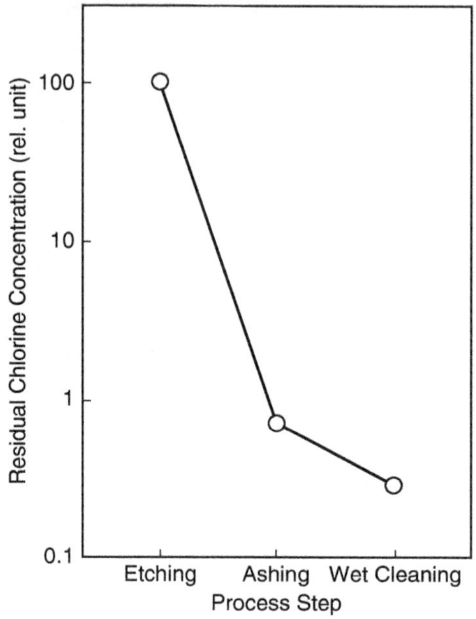

Fig. 3.29 Residual chlorine concentration on the water as a function of process step of Al–Si–Cu etching [27]

necessary to raise the wafer temperature to a fairly high level. There have been reports on etching conducted at 350 °C using $Cl_2 + N_2$ gas chemistry [28], and etching conducted at 280 °C using $SiCl_4 + N_2 + Cl_2 + NH_3$ gas chemistry [29]. However, there would be many additional issues, including the need to use a high-temperature electrostatic chuck and the need to heat all parts of the equipment in order to prevent the deposition of byproducts on the chamber wall and the exhaust system. Because of these issues, etching of the Cu metal lines is not yet used on a commercial basis. Currently, Cu damascene technology, described in Chap. 6, is used for forming the Cu metal lines.

3.4 Summary

The discussions in this chapter have focused on gate etching, SiO_2 etching for holes, spacer etching, and etching of Al alloy stacked metal layer structures, which are the key technologies in the semiconductor manufacturing process. These discussions have delved into details on the key parameters for the etching processes and the methods for controlling them. The semiconductor manufacturing technology continues to advance, and new materials will continue to be introduced in the future. It is important to figure out how to handle these materials by going back to the basics. Readers will hopefully be able to apply the basic concepts of the "mechanism of dry etching" discussed in Chap. 2 and the "dry etching of various materials" discussed in this chapter in their future etching technology development work.

References

1. K. Nojiri, M. Sadaoka, H. Azuma, K. Kawamura: Ext. Abstr. 36th Spring Meeting of The Japan Society of Applied Physics, No. 2, p.571 (1989).
2. M. Nakamura, K. Iizuka and H. Yano: Jpn. J. Appl. Phys., 28, 2142 (1989).
3. S. Ramalingam, Q. Zhong, Y. Yamaguchi and C. Lee: Proc. Symp. Dry Process, p.139 (2004).
4. S. Tachi, M. Izawa, K. Tsujimoto, T. Kure, N. Kofuji, R. Hamasaki, and M. Kojima: J. Vac. Sci. Technol., A16, 250 (1998).
5. C. Lee, Y. Yamaguchi, F. Lin, K. Aoyama, Y. Miyamoto and V. Vahedi: Proc. Symp. Dry Process, p.111 (2003).
6. K. Nojiri, C. Lee, V. Vahedi, R. A. Gottscho: Proc. SEMI Technol. Symp. p.6-14, (2004).
7. K. Nojiri, K. Tsunokuni and K. Yamazaki: J. Vac. Sci. Technol., B14, 1791 (1996).
8. T. P. Chow and A. J. Steckl: J. Electrochem. Soc. 131, 2325 (1984).
9. K. Nojiri, N. Mise, M. Yoshigai, Y. Nishimori, H. Kawakami, T. Umezawa, T. Tokunaga and T. Usui: Proc. Symp. Dry Process, p.93 (1999).
10. H. Horiike: Proc. 19th Semiconductor Technology Seminar, p.193 (1981).
11. L. M. Ephrath: J. Electrochem. Soc. 126, 1419 (1979).
12. M. Izawa, K. Yokogawa, S. Yamamoto, N. Negishi, Y. Momonoi, T, Tsutsumi, K. Tsujimoto, S. Tachi: Ext. Abstr. 46th Spring Meeting of The Japan Society of Applied Physics, No. 2, p.793 (1999).
13. T. Tatsumi, H. Hayashi, S. Morishita, S. Noda, M. Okigawa, N. Itabashi, Y. Hikosaka and M. Inoue: Jpn. J. Appl. Phys. 37, 2394 (1998).
14. S. Tachi, M. Izawa and M. Kojima: Proc. Symp. Dry Process, P83 (1997).
15. S. Ito, K. Nakamura and H. Sugai: Jpn. J. Appl. Phys. 33, L1261 (1994).
16. M. Mori, S. Yamamoto, K. Tsujimoto, K. Yokogawa and S. Tachi: Proc. Symp. Dry Process, p.397 (1997).
17. K. Nojiri and E. Iguchi: J. Vac. Sci. & Technol., B 13, 1451 (1995).
18. N. Negishi, M. Izawa, K. Yokogawa, Y. Momonoi, T. Yoshida, K. Nakaune, H. Kawahara, M. Kojima, K, Tsujimoto and S. Tachi: Proc. Symp. Dry Process, p.31 (2000).
19. T. Tatsumi, M. Matsui, S. Kobayashi, L. Wang, S. Noda and M. Sekine: Proc. Symp. Dry Process, p.297 (1999).
20. T. Sakai, H. Hayashi, J. Abe, K. Horioka and H. Okano: Proc. Symp. Dry Process, p.193 (1993).
21. Y. Ito, A. Koshiishi, R. Shimizu, M. Hagiwara, K. Inazawa and E. Nishimura: Proc. Symp. Dry Process, p.263 (1998).
22. I. Hasegawa, Y. Yoshida, Y. Naruke and T. Watanabe: Proc. Symp. Dry Process, p.126 (1985).
23. Y. Kawasaki, M. Nishiumi, S. Okudaira, Y. Kakehi: Hitachi Review, Vol 71, No. 5, p.33 (1989).
24. M. Sadaoka, K. Nojiri, K. Hirobe, R. Fukuyama: Ext. Abstr. 48th Fall Meeting of The Japan Society of Applied Physics, No. 2, p.469 (1987).
25. N. Tamura, K. Nojiri, K. Tsujimoto, K. Takahashi: Hitachi Review, 44 (2) 91 (1995).
26. A. Hall and K. Nojiri: Solid State Technol., 34 (5) 107 (1991).
27. K. Nojiri: 0.3μm Process Technology, p.163 (TRICEPS INC, 1994).
28. K. Hoshino, M. Nakamura, H. Yagi, H. Yano, H Tsuchikawa: Ext. Abstr. 36th Spring Meeting of The Japan Society of Applied Physics, No. 2, p.570 (1989).
29. K. Ohno, M. Sato and Y. Arita: Ext. Abstr. Int. Conf. Solid State Devices and Materials, p.215 (1990).

Chapter 4
Dry Etching Equipment

This chapter first reviews the history of dry etching equipment. Then it provides detailed discussions of the basic mechanisms of plasma generation, plasma density, operating pressure conditions, and key characteristics of dry etching equipment used in LSI manufacturing today, including the barrel-type plasma etcher, capacitively coupled plasma (CCP) etcher, magnetron reactive-ion etching (RIE), electron-cyclotron resonance (ECR) plasma etcher, and inductively coupled plasma (ICP) etcher. Finally, the electrostatic chuck, which plays an important role in dry etching equipment, is discussed.

4.1 History of Dry Etching Equipment

The dry etching equipment first used in semiconductor manufacturing as an alternative to wet etching was a barrel-type plasma etcher. In this system, plasma is generated with an electromagnetic induction coil or capacitively coupled electrodes placed around a cylindrical quartz tube. It was used on semiconductor materials for the first time in 1968, but applications were limited to process steps that did not require high-precision processing, such as resist ashing, wafer backside film removal, and etching of the dielectric films on bonding pads, because the processing precision was not very high.

Dry etching technology came to be used in earnest for semiconductor manufacturing starting with the 3-μm process for 64K-bit DRAM. The equipment used was the parallel-plate dry etching system called CCP, which generated plasma by capacitive coupling. This type of etching equipment was also referred to as RIE and became the most widely known type of dry etching system. Many improvements have been made on RIE technology since then for a wide range of uses on poly–Si, dielectric films, Al lines, and other materials. A batch system was used in the early days in order to process a large number of wafers at once. As wafers became larger in diameter, and more precise processing was required, the batch system no longer

K. Nojiri, *Dry Etching Technology for Semiconductors*,
DOI 10.1007/978-3-319-10295-5_4

worked well enough, and single-wafer dry etching equipment, for processing wafers one at a time, was developed. With single-wafer etching equipment, it was necessary to increase the etch rate in order to achieve throughputs that should be comparable to those of the batch system, and achieving higher plasma densities became an important technological challenge.

The first method utilized for increasing the plasma density in single-wafer dry etching equipment was the magnetron RIE. This is a method of enhancing the plasma density with the application of a magnetic field and is also referred to as MERIE, for magnetically enhanced reactive-ion etching. This method has little margins in terms of wafer charging resulting from nonuniform plasma. Because of this problem, applications for this technology became limited as devices were scaled more. Furthermore, the plasma density is lower than that of the ECR and ICP technologies to be discussed later.

An ECR plasma etcher, which generates a high-density plasma through ECR resulting from mutual interaction between microwave and magnetic field, was introduced at the 0.8-μm technology node for 4M-bit DRAM. Hitachi commercialized this technology for volume production equipment and launched it commercially in 1985. ECR plasma etching has advantageous features to realize fine patterning, such as high-density plasma generation at a low-operating pressure, and independent control of ion energy and plasma discharge. It became the dominant single-wafer, high-density plasma etcher for gate and Al line etching. While Hitachi calls this technology microwave plasma etching, it is referred to as ECR plasma etching in this book because the technology does fall into this category.

An ECR plasma etcher requires a large magnetic coil to form a strong magnetic field, which makes it difficult to make the etch chamber compact. The ICP etcher was developed to address this issue. A simple coil was placed on top of the etch chamber to generate a high-density plasma by electromagnetic induction. The ICP etcher gradually established itself as the high-density-plasma, single-wafer etching equipment of choice because it offered a plasma density comparable to that of an ECR plasma etcher without using a large electromagnetic coil and, like the ECR plasma etcher, offered an ability to control the ion energy independently of the plasma discharge. An important example of an ICP etcher was the transformer-coupled plasma (TCP®) etcher launched by Lam Research in 1992.

High-density-plasma technologies such as ECR and ICP are used today for fine patterning of gate materials and conductive materials like Si and Al. On the other hand, as discussed in Chap. 3, high-density plasma is not suited for SiO_2 etching. Instead, as discussed earlier, a narrow-gap, parallel-plate CCP etcher with a medium plasma density and 20–30-mm gap and the magnetron RIE are used for SiO_2 etching.

In order to address many issues arising with larger wafer diameters and finer geometries, the dry etching technology that began with batch-type CCP has gone through many changes until today. The transition from batch to single-wafer etcher was a necessity for processing larger-diameter wafers. In order to realize this transition, a higher plasma density to increase the etch rate was required. At the same time, processing of fine patterns requires operation at a lower pressure. Therefore, the history of dry etching technology was the history of innovations for attaining a higher plasma density at lower pressures.

The subsequent sections in this chapter include detailed discussions of the dry etching equipment used for LSI manufacturing.

4.2 Barrel-Type Plasma Etcher

Figures 4.1 and 4.2 show the cross-sectional diagrams for the barrel-type plasma etchers. This system has an electromagnetic induction coil (Fig. 4.1) or a pair of capacitively coupled electrodes (Fig. 4.2) surrounding a cylindrical quartz tube. Plasma is generated with an application of the 13.56-MHz RF power. The equipment shown in Fig. 4.1 generates the plasma by induction coupling, while the equipment in Fig. 4.2 generates the plasma by capacitive coupling. Approximately 50 wafers are placed inside the quartz tube, and these wafers are processed in a batch. The operating pressure is 10–10^3 Pa, and etching proceeds isotropically by radical reactions. While the barrel-type plasma etchers were the first equipment used for the semiconductor manufacturing process as an alternative to wet etching, due to isotropic etching they were used on steps that did not require high-precision processing, such as resist ashing, wafer backside film removal, and etching of the dielectric films on bonding pads. They are mainly used for resist ashing today

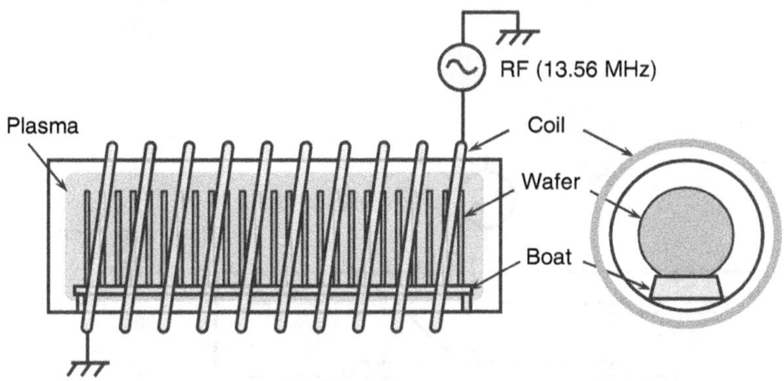

Fig. 4.1 Schematic diagram of barrel-type plasma etching apparatus (inductively coupled)

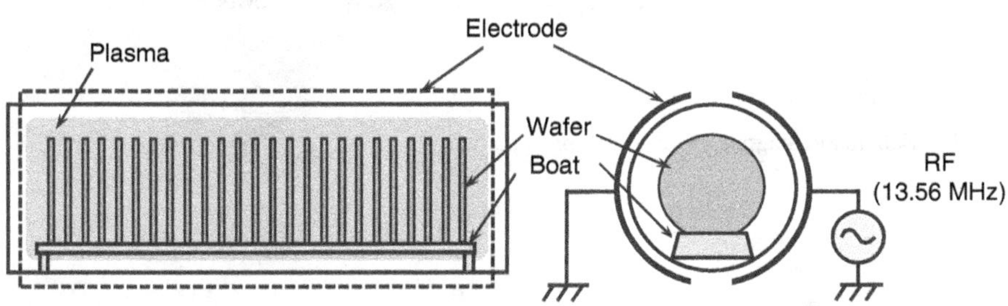

Fig. 4.2 Schematic diagram of barrel-type plasma etching apparatus (capacitively coupled)

4.3 Capacitively Coupled Plasma Etcher

Figure 4.3 shows a cross-sectional diagram of a CCP etcher. Plasma is generated by applying RF power on a pair of electrodes that are placed in parallel with each other. This type of etcher is also referred to as a parallel-plate plasma etcher, or an RIE, and was the first type of equipment that was widely used for an actual patterning process. In the early days, it was used for materials such as poly–Si, dielectric film, and Al lines. Subsequently, high-density plasma technologies such as ECR and ICP came to be widely used for poly–Si and Al line etching. CCP etchers are still in wide use today for SiO_2 etching. A narrow-gap, parallel-plate etcher, with the distance between the electrodes narrowed down to 20–30 mm, is used for SiO_2 etching. While relatively higher operating pressures, of around 100–200 Pa, were used in the past, the pressure levels were later reduced to improve the processing precision. The pressure regime used today is around 1–5 Pa. The plasma density is on the order of 10^{10} cm^{-3}. An RF power at two frequency levels (the high and low frequencies) is usually applied. The high frequency is usually at 27–60 MHz, while the low frequency is usually at 800 kHz to 2 MHz. High-frequency power is mainly used for generating the plasma, while low-frequency power is used for controlling the ion energy. Equipment from Lam Research and Tokyo Electron are representative examples of narrow-gap, parallel-plate etchers. Figure 4.4 shows a cross-sectional diagram of the 2300® Exelan® equipment from Lam Research [1].

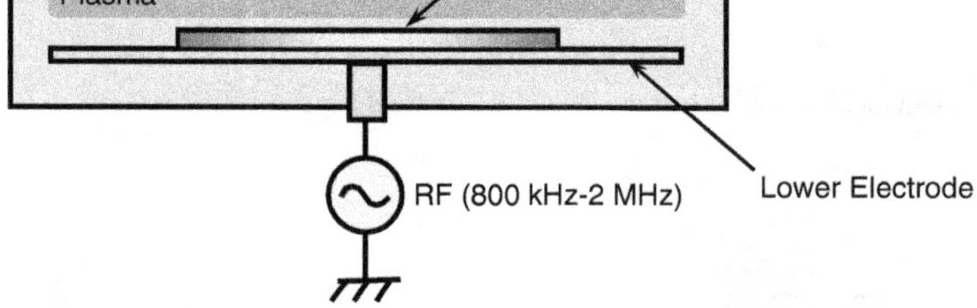

Fig. 4.3 Schematic diagram of CCP etching apparatus

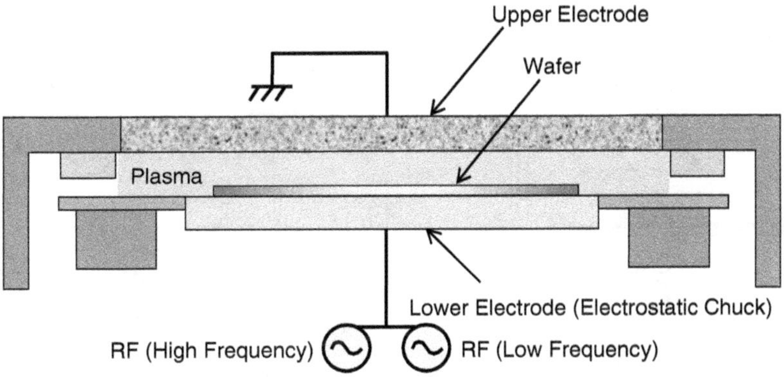

Fig. 4.4 Lam Research 2300® Exelan® [1]

4.4 Magnetron Reactive-Ion Etching

Magnetron RIE is a method of generating a high-density plasma through the cycloid motion of electrons resulting from mutual interactions between the electrical and the magnetic fields. The equipment structure is a parallel-plate RIE on which a magnetic field, generated by a permanent magnet or an electromagnetic coil, is applied. Tokyo Electron's dipole-ring magnet (DRM) is a representative example of the magnetron RIE equipment. A cross-sectional diagram is shown in Fig. 4.5 [2]. Figure 4.6 illustrates the principle of magnetron discharge. When a magnetic field (B) is applied in a direction orthogonal to the sheath electrical field (E) or in a direction parallel to the lower electrode, the electrons are subjected to the Lorentz force ($E \times B$) and travel in a direction orthogonal to both the electric field and the magnetic field along a cycloid curve. This results in an increased probability of collisions, and a high-density plasma is thus generated. The magnetron RIE is operated at a pressure of around 1 Pa, and the plasma density is on the order of 10^{10} cm^{-3}.

The magnetron technology is capable of generating a high-density plasma even at a low pressure. However, it has some drawbacks. First, ion energy cannot be controlled independently of the plasma discharge, because the plasma discharge and the ion energy control are carried out with a single RF power supply. Second, it is extremely difficult to form a uniform plasma, because the electrons tend to travel in one direction in a cycloid motion. The second drawback leads to problems such as an etch-rate nonuniformity and gate oxide breakdown due to wafer charging [3]. It is possible to make the etch rate more uniform with a scanning or rotating motion of the magnet. However, when the gate oxide breakdown occurs under a static magnetic field, it cannot be prevented even if the magnetic field is rotated. It has been reported that the gate oxide breakdown in the magnetron RIE can be prevented by optimizing the distribution of the magnetic field [4]. However, it becomes increasingly difficult to obtain a uniform plasma, as the wafer diameter increases, and this technology is being used on fewer etch process steps these days.

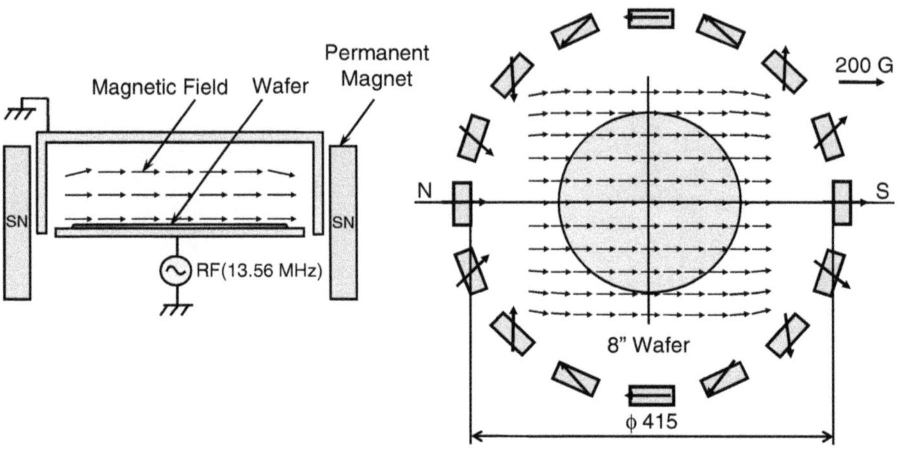

Fig. 4.5 Schematic diagram of dipole-ring magnet (DRM) [2]

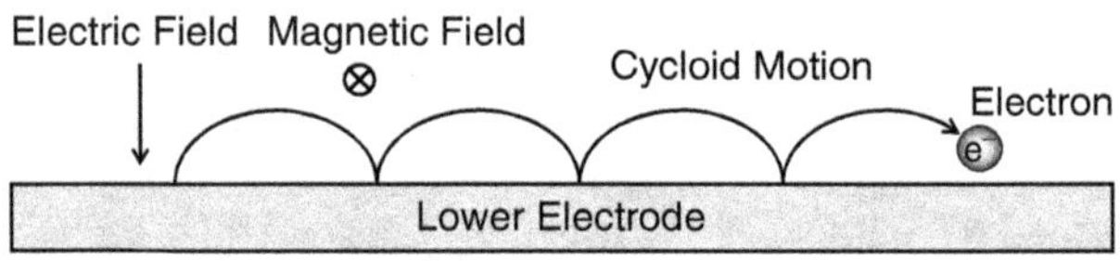

Fig. 4.6 Principle of magnetron discharge

4.5 Electron-Cyclotron Resonance Plasma Etcher

An ECR plasma etcher has advantageous features, such as high-density plasma generation at a low pressure and independent control of ion energy and plasma discharge. These features are suited for fine pattern etching. The production system that Hitachi commercialized became the dominant single-wafer, high-density plasma etcher in a certain period, for gate and Al line etching.

Figure 4.7 shows the structure of the etch chamber of an ECR plasma etcher [5], while Fig. 4.8 shows how the ECR plasma is generated. The 2.45-GHz microwave generated by a magnetron travels through a wave guide and is introduced into the etch chamber through a quartz window. An electromagnetic coil is placed around the chamber, and the electrons go into a cyclotron motion under the electric field of the microwave and a magnetic field that is formed orthogonally to the electric field. When the microwave frequency is 2.45 GHz, a magnetic field of 875 G would result in an ECR, which in turn increases the probability of collisions and makes it possible to generate a high-density plasma at a low pressure. The operating pressure is around 1 Pa, and the plasma obtained at this pressure is of very high density at 10^{11} cm^{-3} or above. It is possible to control the ion energy independently of the

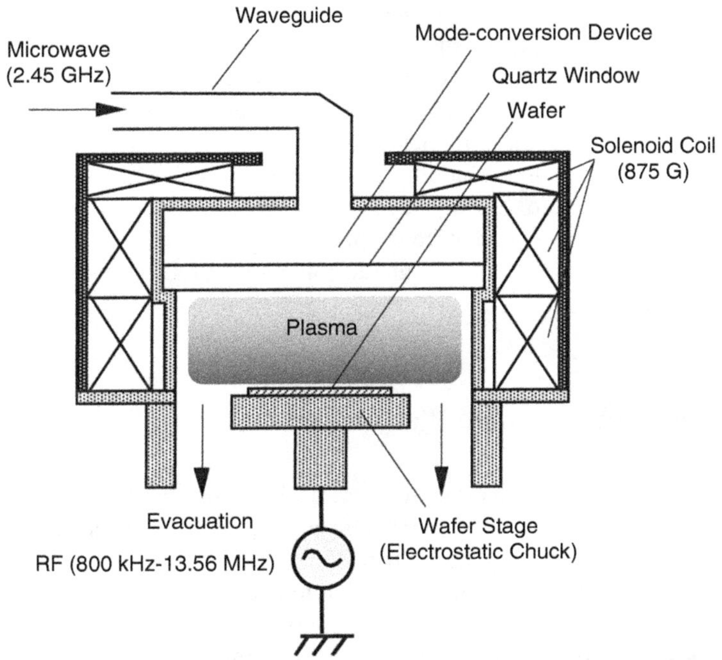

Fig. 4.7 Schematic diagram of ECR plasma etching apparatus [5]

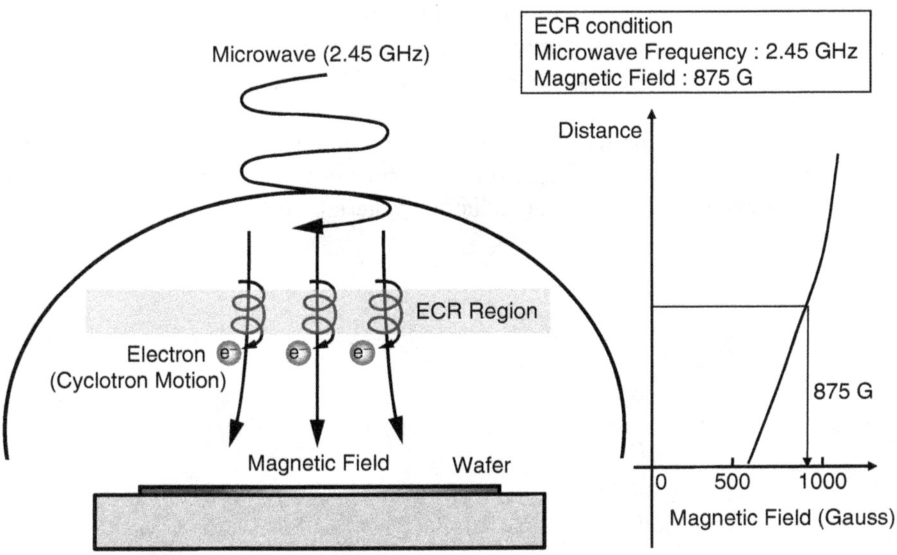

Fig. 4.8 Principle of ECR plasma generation

plasma discharge through an RF power applied to the wafer stage, and a closely controlled fine patterning is possible. Because the ECR plasma uses the magnetic field, however, charging damage occurs under some conditions. This issue can be resolved by lowering the frequency of the RF generator connected to the wafer stage [3]. Details are discussed in the next chapter.

Hitachi later went on to develop the UHF-ECR plasma etcher. In this system, a UHF band, which is approximately one fifth the frequency of a microwave, is used, and the magnetic field is also reduced to approximately one fifth [6]. The equipment was made smaller, with a smaller magnetic coil, and the risks of charging damages were also reduced.

4.6 Inductively Coupled Plasma Etcher

The ICP dry etching equipment includes the TCP® etcher from Lam Research and decoupled plasma source (DPS®) etcher from Applied Materials [8]. Figure 4.9 shows a cross-sectional diagram for the TCP® etcher [7], while Fig. 4.10 shows how the TCP® plasma is generated. An induction coil (TCP® coil) is placed on top of an insulating plate (TCP® window) at the top of the etch chamber, and a 13.56-MHz RF power supply (source power) for generating the plasma is connected to the coil. A magnetic field is created when a high-frequency current flows through the TCP® coil. This magnetic field, in turn, creates an electric field in the chamber and generates a high-density plasma. A 13.56-MHz RF power supply for controlling the ion energy (bottom power) is connected to the wafer stage, and the ion energy is controlled independently of the plasma discharge. The operating pressure is around 1 Pa, and a high-density plasma of 10^{11} cm^{-3} or more is obtained at this pressure regime.

Because the ICP etcher is able to provide a high-density plasma without a large electromagnetic coil as in an ECR plasma etcher, it has become the mainstream etching equipment for processing conductive materials such as gates, Si (e.g., STI), and Al lines.

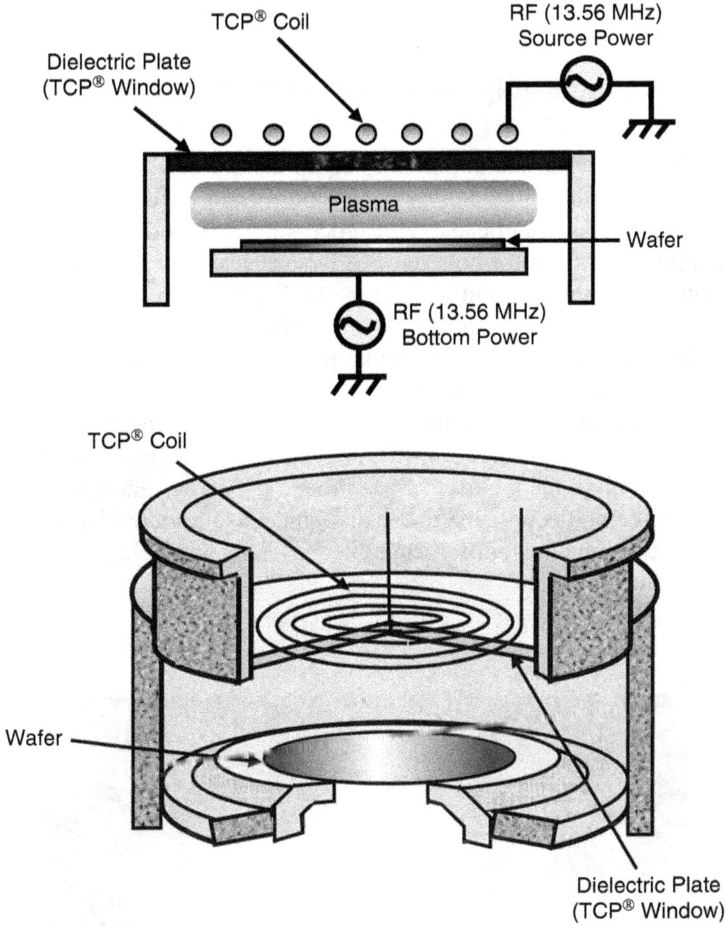

Fig. 4.9 Schematic diagram of TCP® etching apparatus [7]

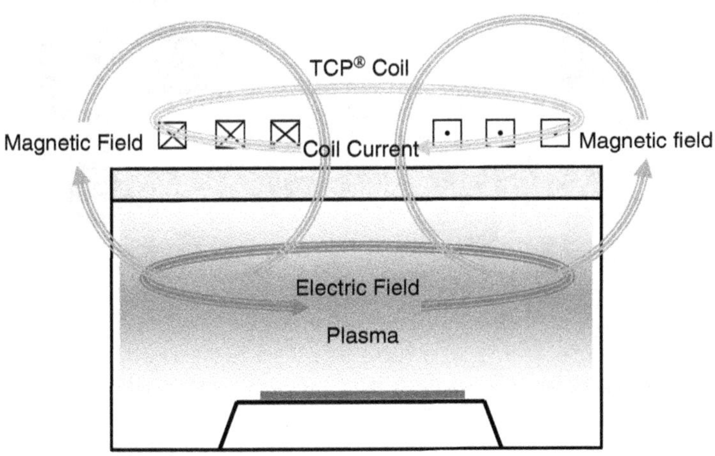

Fig. 4.10 Principle of TCP® plasma generation

4.7 Dry Etching Equipment for Manufacturing

As a real-life example of the dry etching equipment, Fig. 4.11 shows an external view photograph of the 2300® Series system from Lam Research. It is the single-wafer dry etching equipment capable of processing 300- and 200-mm wafers and offers a common platform for both the TCP® plasma source, used for etching conductive materials such as Si and Al, and the CCP plasma source, used for etching dielectric film like SiO_2. At the right side of the photograph is the equipment's front side, with the wafer loading and unloading part. The etch chambers are to the left rear of the photograph. In order to increase the throughput, the dry etching equipment used for semiconductor manufacturing today generally includes a multitude of etch chambers on a single platform in a multi-chamber architecture. Figure 4.12 shows a photograph of the multi-chamber structure. This system can accommodate up to four chambers. It is possible to use the same plasma source for all four chambers or to use combinations of, for example, the TCP® and Asher or the TCP® and CCP on the same platform.

Fig. 4.11 External view photograph of dry etching system (Lam Research 2300® Series)

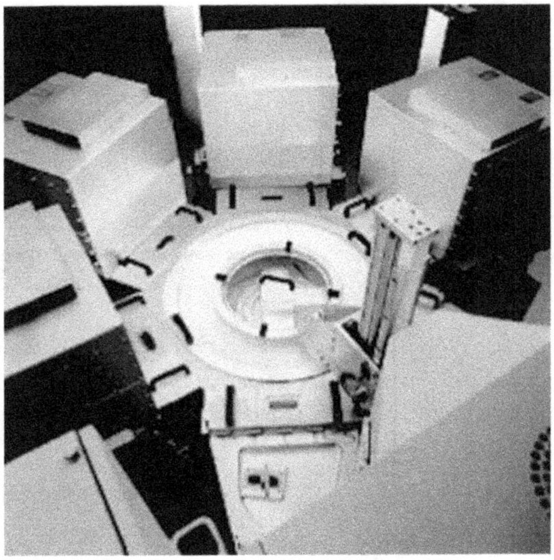

Fig. 4.12 Photograph of multi-chamber structure (Lam Research 2300® Series)

4.8 Electrostatic Chuck

Finally, we will discuss the electrostatic chuck, which plays an important role in dry etching equipment. The electrostatic chuck is sometimes referred to as the ESC. In etching equipment today, the electrostatic chuck is usually used as the lower electrode. As etching proceeds, the wafer is heated by the plasma, and its temperature rises. As mentioned earlier in Sect. 3.1 in Chap. 3, the etch profile and CD are greatly affected by the redeposition of reaction byproducts onto the pattern sidewalls. Because the sticking probability of the reaction byproducts is strongly dependent on temperature, it is necessary to control the wafer temperature for high-precision etching. The electrostatic chuck relies on electrostatic force for wafer clamping and maintains the wafer temperature at a constant level during etching. Mechanical clamps were used in the earlier days for clamping the wafer to the electrode. The mechanical clamps fasten the wafer periphery to hold it against the electrode. For this reason, contact at the wafer center tends to be inadequate, and the areas covered by the clamps do not get etched. The electrostatic chuck avoids these issues and has become a widely used method of wafer clamping today.

4.8.1 Types of Electrostatic Chucks and Chucking Mechanisms

An electrostatic chuck may be categorized as a monopolar or bipolar type, depending on its structure. Figure 4.13 shows the cross-sectional structures and clamping mechanisms. An electrostatic chuck basically clamps a wafer to the electrode through a Coulomb force between the wafer and the electrode, which is under an attractive

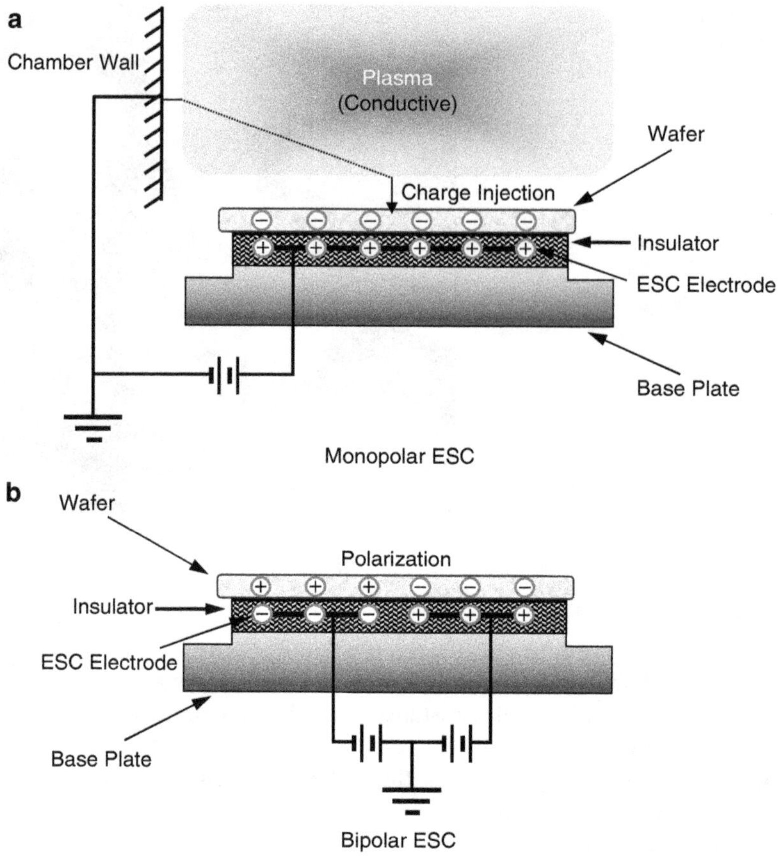

Fig. 4.13 Cross-sectional structures and clamping mechanisms for monopolar and bipolar electrostatic chucks

force between positive and negative charges. With the monopolar structure, the wafer receives electrical charges from the plasma for clamping. In other words, there is no clamping without a plasma. Conversely, with the bipolar structure, the wafer becomes polarized internally, and clamping without the plasma is possible.

Furthermore, the electrostatic chuck may be categorized into either the Coulombic electrostatic chuck or the Johnsen–Rahbek electrostatic chuck, based on the differences in the conductivity of the materials. Figures 4.14a and b show their respective cross-sectional diagrams. With the Coulombic electrostatic chuck, the wafer and an ESC electrode are isolated by an insulator (resistivity $>10^{15}$ Ωcm), and no charges travel through. The insulating material may be an alumina ceramic or a polyimide.

When a high voltage of around 3,000 V is applied on the ESC electrode, charges with opposite polarity are induced at the wafer backside surface, and the wafer is clamped as a result of the Coulomb force between these charges. Because the Coulombic electrostatic chuck involves no movement of charges, it offers fast

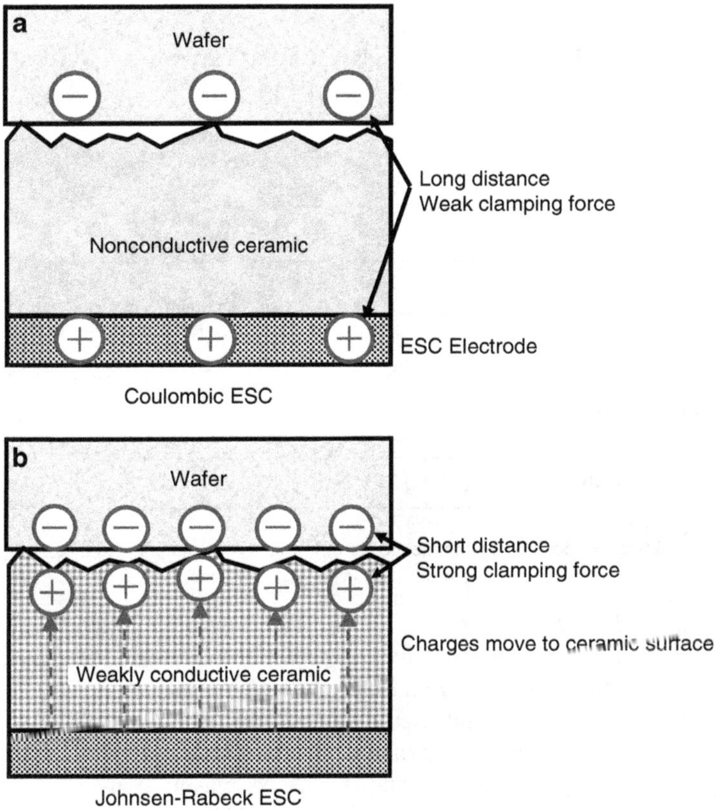

Fig. 4.14 Cross-sectional structures and clamping mechanisms for Coulombic and Johnsen–Rabeck electrostatic chucks

chucking and dechucking responses. However, the clamping force is small, and a high voltage is required for the clamping. The resistivity is not dependent on temperature.

With the Johnsen–Rahbek electrostatic chuck, a material of some level of electrical conductivity (a resistivity greater than 10^9–10^{12} Ωcm) is inserted between the wafer and the electrostatic chuck electrode. This material is usually an alumina ceramic doped with impurities, such as TiO_2. When a voltage is applied on the electrostatic chuck electrode, the charges move through the ceramic and collect at the surface. For this reason, the distance between the positive and negative charges is smaller, and the clamping force is larger. On the other hand, it takes time for the charges to move, and the chucking and dechucking response is slower than that of a Coulombic electrostatic chuck. Furthermore, the resistivity is dependent on temperature.

Figure 4.15 shows examples of a chucking and dechucking sequence for a bipolar-type Johnsen–Rahbek electrostatic chuck. The solid line represents the voltage applied on the positive electrode, while the broken line represents the voltage applied on the negative electrode. Charges move to the ceramic surface when a

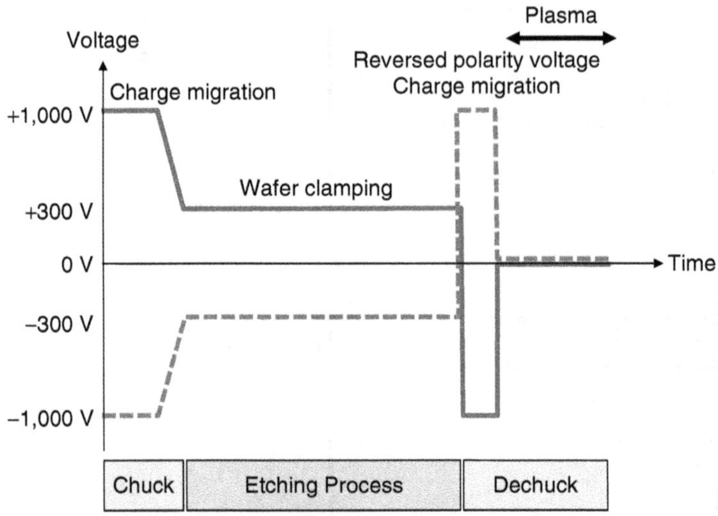

Fig. 4.15 Chucking and dechucking sequence for a bipolar-type Johnsen–Rahbek ESC

voltage of around 1,000 V is applied on the electrodes, and these charges in turn induce charges of an opposite polarity on the wafer backside. These positive and negative charges attract each other and keep the wafer clamped. During the etching process, wafer clamping is maintained with a voltage of around 300 V. The wafer is dechucked when a reversed polarity voltage of around 1,000 V is applied for removing the charges from the ceramic surface. A plasma is usually used in the dechucking sequence in order to remove the residual charges.

4.8.2 Principle of Wafer Temperature Control

The temperature of the wafer is controlled indirectly through a thermal contact between the temperature-controlled lower electrode (electrostatic chuck) and the wafer backside. Figure 4.16 shows the principle of wafer temperature control. The electrostatic chuck is maintained at a specific temperature by a coolant circulating from the chiller. Since the physical contact between the electrostatic chuck and the wafer does not ensure enough transfer of heat, the space between the wafer and the electrostatic chuck is filled with He gas to promote thermal conduction. Helium is lighter than air and the etching gas, and molecules move back and forth rapidly between the wafer and the lower electrode to transport the thermal energy. Because the thermal conductivity of He is approximately six times higher than that of air or etching gas, the heat is transferred efficiently.

The etching profile and CD are strongly affected by the redeposition of reaction byproducts on pattern sidewalls. The sticking coefficient of the reaction byproducts strongly depends on temperature. Therefore, it is important to control the temperature

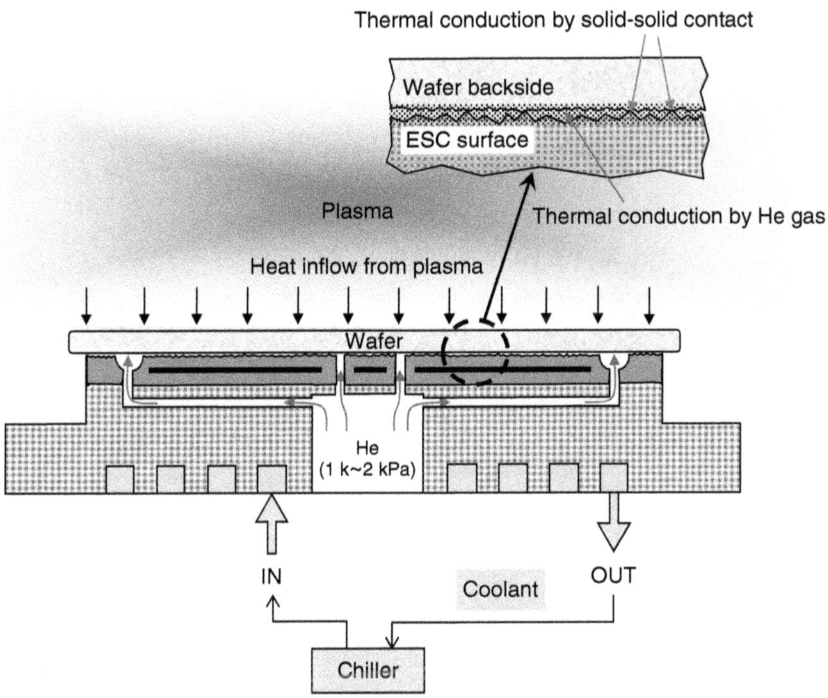

Fig. 4.16 Principle of water temperature control

distribution across the wafer in order to improve the across-the-wafer uniformity of the CD and etching profiles. For this reason, an electrostatic chuck capable of independent control of temperature at the center and at the periphery of the wafer has been developed in recent years, as described in Sect. 3.1.2 in Chap. 3 [9].

References

1. K. Nojiri: Electronic Journal 90th Technical Symposium, p.89 (2004).
2. M. Sekine, M. Narita, S. Shimonishi, I. Sakai, K. Tomioka, K. Horioka, Y. Yoshida and H. Okano: Proc. Symp. Dry Process, p.17 (1993).
3. K. Nojiri and K. Tsunokuni: J. Vac Sci. Technol., B 11, 1819 (1993).
4. S. Nakagawa, T. Sasaki, and M. Mori: Proc. Symp. Dry Process, p.23 (1993).
5. K. Nojiri, N. Mise, M. Yoshigai, Y. Nishimori, H. Kawasaki, T. Umezawa, T. Tokunaga, and T. Usui: Proc. Symp. Dry Process, p.93 (1999).
6. N. Negishi, M. Izawa, K. Yokogawa, Y. Morimoto, T. Yoshida, K. Nakamura, H. Kawahara, M. Kojima, K. Tsujimoto, and S. Tachi: Proc. Symp. Dry Process, p.31 (2000).
7. J. B. Carter, J. P. Holland, E. Pelzer, B. Richardson, E. Bogle, H. T. Nguyen, Y. Melaku, D. Gates, and M. Ben-Dor: J. Vac Sci. Technol., A 11, 1301 (1993).
8. T. Fukamachi: Electronic Journal 24th Technical Symposium, p.81 (1999).
9. C. Lee, Y. Yamaguchi, F. Lin, K. Aoyama, Y. Miyamoto, and V. Vahedi: Proc. Symp. Dry Process, p.111 (2003).

Chapter 5
Dry Etching Damage

Dry etching has been a key technology in the LSI manufacturing process, and the high integration of LSI would not have been realized without progress in this technology. However, because the process uses plasma, the devices are susceptible to various types of damages caused by high-energy and charged particles. A large amount of damage sometimes lowers the LSI yields and reliability.

As Fig. 5.1 shows, damage introduced by dry etching includes (1) the creation of crystal defects and invasion of impurities caused by the ionic impacts on the Si surface layer, (2) wafer charging caused by charged particles, and (3) neutral traps formed by high-energy photons. Some of the problems that become serious as a result of device scaling are charge retention failure in DRAM and increased contact resistance caused by damage on the Si surface, and gate oxide breakdown caused by wafer charging. This chapter reviews these types of damage, which have grave impacts on the devices, and discusses their causes and potential solutions. The effects of these kinds of damage on device characteristics are also discuss.

K. Nojiri, *Dry Etching Technology for Semiconductors*,
DOI 10.1007/978-3-319-10295-5_5

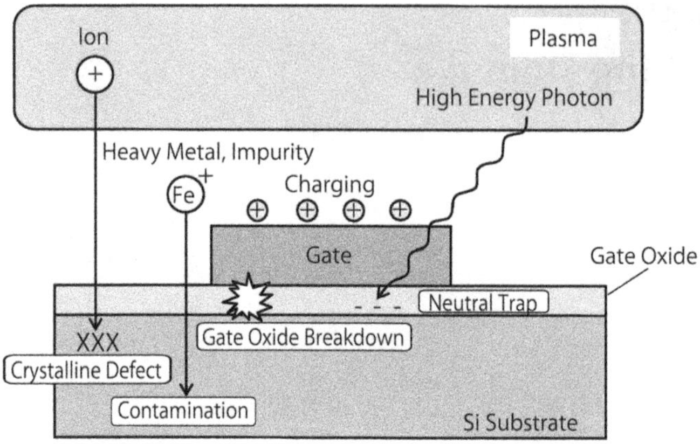

Fig. 5.1 Various damage induced by dry etching

5.1 Damage Induced in the Silicon Surface Layer

During dry etching, impurities such as heavy metals are released into the plasma from the etch chamber or the electrode, either by sputtering or by chemical reactions. These impurities or molecules in the etching gas are injected into the Si substrate in the form of ions. In SiO_2 etching using fluorocarbon-based gases, C and F remain within an approximately 10-nm range from the surface [1], but H penetrates deeper. It has been reported that H was present at a depth of 50 nm with a reactive-ion etching (RIE) using CF_4+H_2 gas chemistry [2]. Figure 5.2 shows the result of a high-resolution transmission electron microscope analysis on the Si surface layer that has been damaged in a narrow-gap parallel-plate etcher [3]. The peak-to-peak voltage (V_{pp}) during etching is 2.5 kV. There is a layer of crystal disorder down to a depth of approximately 2.5 nm from the surface. Also, crystal defects are present down to a depth of 50 nm. Because of this depth distribution, one may surmise that the crystal defects are related to H penetrating into the Si crystal. Figure 5.3 shows the result of an atomic force microscopy (AFM) observation on the surface of the sample [3]. While the surface roughness of a sample without plasma treatment is around 0.5 nm, the surface roughness of a sample with plasma treatment is around 2.5 nm. Heavy metal contamination and crystal defects cause charge retention failure in the DRAM and increased junction leakage current. They are mainly caused by the generation recombination (GR) centers formed in the Si substrate by the heavy metal contamination. A solution is, first, to remove the sources of heavy metal contamination inside the etch chamber. In other words, materials that could turn into sources of heavy metal contamination, such as stainless steel, should not be used for areas that contact the plasma. It is also necessary to use high-purity materials, even if quartz or Al is utilized in the etch chamber. Next, an effective approach is to reduce the ion energy. Figure 5.4 shows the minority carrier lifetime in electron-cyclotron resistance (ECR) plasma etching of SiO_2, as a function of the

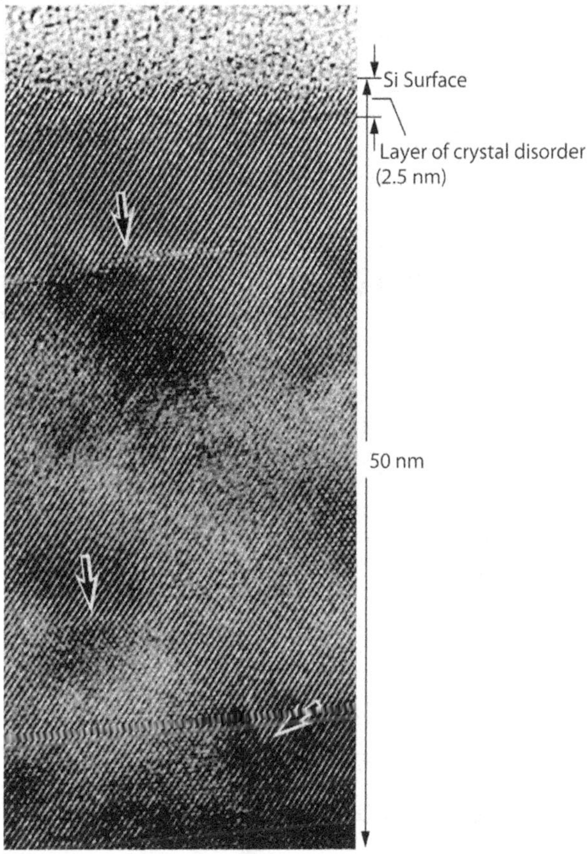

Fig. 5.2 High-resolution transmission electron microscope photograph of the Si surface layer damaged in a narrow-gap parallel-plate plasma etcher. *Arrows* indicate crystalline defects [3]

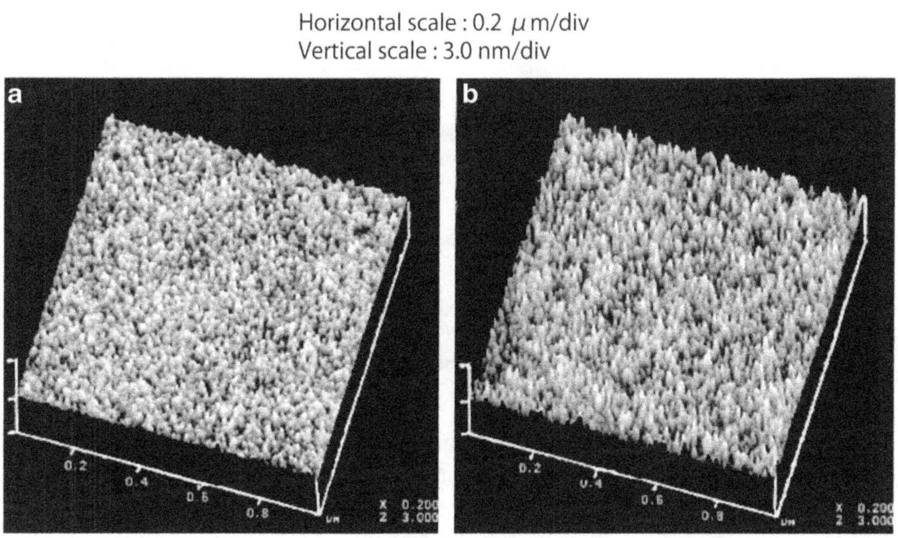

Fig. 5.3 Atomic force microscopy image for the samples with and without plasma treatment [3]

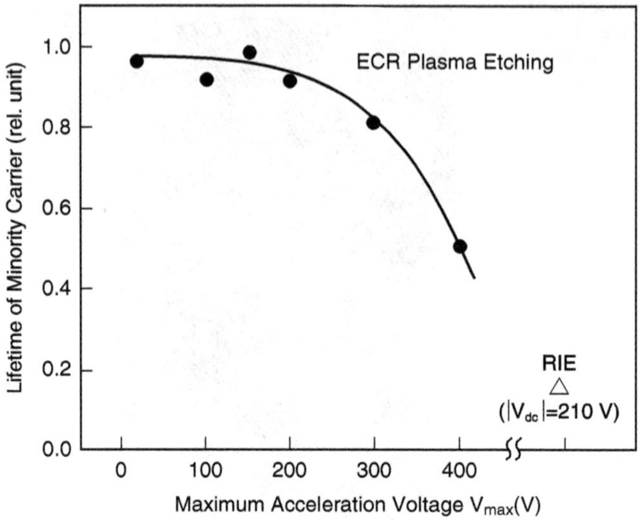

Fig. 5.4 Lifetime of minority carrier in ECR etching as a function of maximum acceleration voltage (V_{max}) [4]

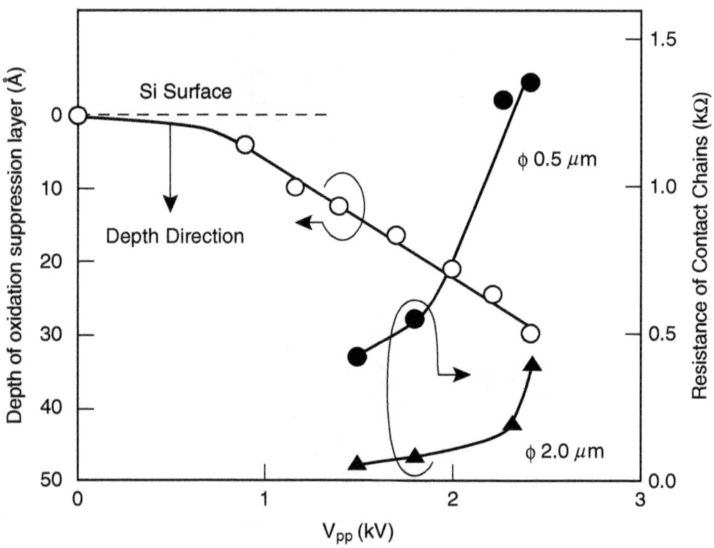

Fig. 5.5 Depth of oxidation suppression layer and contact resistance as a function of V_{pp} [5]

maximum acceleration voltage for the ions (V_{max}) [4]. The figure shows that the degradation of the lifetime is suppressed up to $V_{max} = 300$ V.

As devices are scaled, an increase in contact resistance becomes a more significant problem. The effects of H used to be considered the main factor to cause the increase in resistance, but the Si–C [5, 6] and Si–O layers [7], which are formed near the surface, are now considered the main source of the problem. Figure 5.5

shows the V_{pp} dependence of contact resistance when SiO_2 etching is conducted on a narrow-gap parallel-plate etcher [5]. When V_{pp} exceeds 2 kV, the contact resistance begins to increase rapidly. At the same time, an increase in the thickness of an oxide suppression layer consisting of Si–C is observed and is considered the cause of the increased contact resistance. A solution to this problem is to reduce the ion energy. For example, it has been reported that the formation of the Si–C layer can be suppressed by using a ECR plasma etcher capable of low-ion-energy etching [6].

The damaged layer created in the Si surface layer can be removed, for example, with a down-flow–type etcher using O_2+CF_4. However, because the diffusion layers continue to become shallower, there is no longer room to strip away the surface layer. Therefore, a fundamental solution would be to keep the damaged layer as thin as possible during etching. An effective way to achieve that objective is to reduce the ion energy immediately before the Si substrate is exposed.

5.2 Charging Damage

Gate oxide breakdown caused by wafer charging is becoming a serious problem as the gate oxides become thinner. Figure 5.6 shows the frequency of gate oxide breakdown as a function of the gate oxide thickness when a negative wafer charging occurs in a barrel asher [8]. The parameter ΔV_{FB} is a flatband voltage shift of the metal nitride oxide Si (MNOS) capacitors, to be discussed later, and corresponds to the magnitude of the wafer charging. The figure shows that the gate oxide breakdown occurs more frequently with thinner gate oxides. When $\Delta V_{FB}=-2.6$ to -2.4 V, more than 90% of the gates with a oxide film thickness of 4 nm are broken. When developing a dry etching process, it is necessary to monitor the wafer charging and set the conditions and/or improve the equipment and process in order to prevent the gate oxide breakdown.

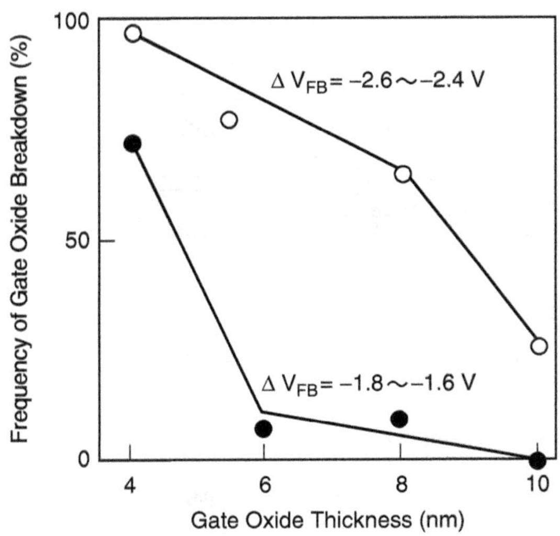

Fig. 5.6 Gate oxide thickness dependence of the gate oxide breakdown frequency for the wafer treated with the barrel asher [8]

5.2.1 Evaluation Method of Charging Damage

The wafer charging resulting from a plasma treatment can be monitored using MNOS capacitors [9]. Figure 5.7a shows the structure of an MNOS capacitor [10]. When this MNOS capacitor is exposed to a plasma, the gate electrode is charged, and depending on the polarity, either the electrons or the holes are injected from the substrate and trapped in the Si_3N_4. As a result, the flatband voltage V_{FB} of the MNOS capacitor shifts. Therefore, the polarity and the amount of charge can be evaluated by measuring the $C-V$ curve and determining the flatband voltage shift ΔV_{FB} of the MNOS capacitor, as shown in Fig. 5.7b. As the example in Fig. 5.7 shows, when the gate is charged positively, ΔV_{FB} is positive. It should be noted that the potential and polarity measured by the MNOS capacitors are relatively defined with respect to the Si substrate. This is explained in the discussion of the potential distribution model (Fig. 5.10) in the next section. The antenna MOS capacitor shown in Fig. 5.8 is generally used for evaluating the gate oxide breakdown [10]. In this structure, an antenna (poly–Si) that collects charges is connected to a gate. This antenna increases the sensitivity of measurement. Here, the antenna ratio r is defined as r=antenna area (S_a)/ gate area (S_g). Figure 5.9 shows how the gate oxide breakdown is dependent on the antenna ratio [10]. Even when the gate oxide breakdown does not occur with an antenna ratio of 1, the breakdown frequency increases with an increased antenna ratio.

5.2.2 Mechanism of Wafer Charging

When the distribution of the RF current flowing onto the wafer surface is nonuniform across the wafer, wafer charging occurs [10]. For example, when the RF current flowing onto the wafer is higher at the wafer peripheral than at the center,

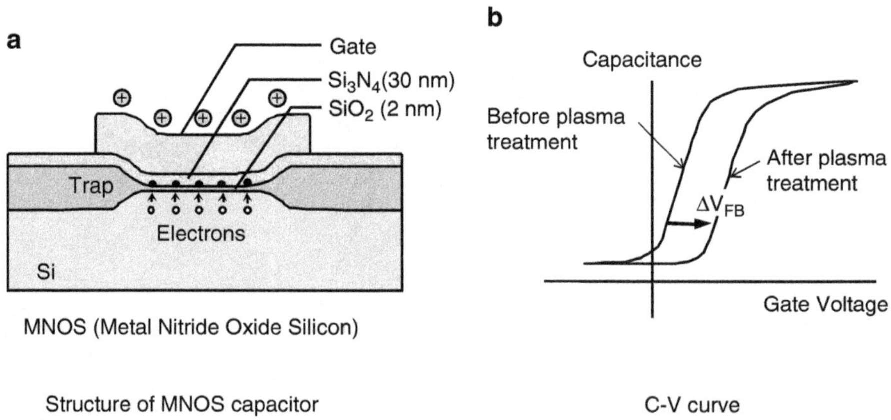

Fig. 5.7 Structure of the MNOS capacitor and the method for charging measurement [10]

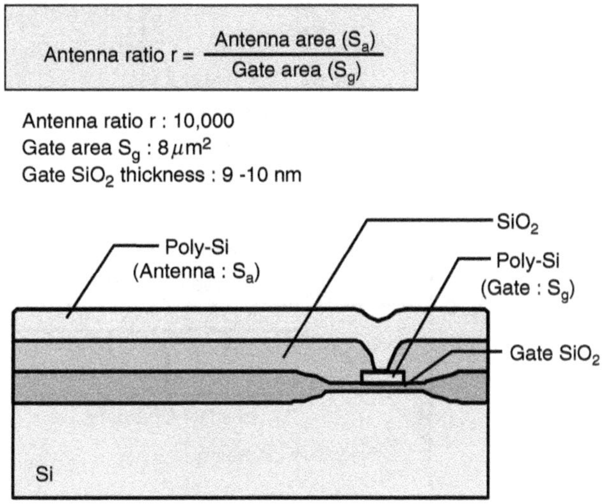

Fig. 5.8 Structure of antenna MOS capacitor [10]

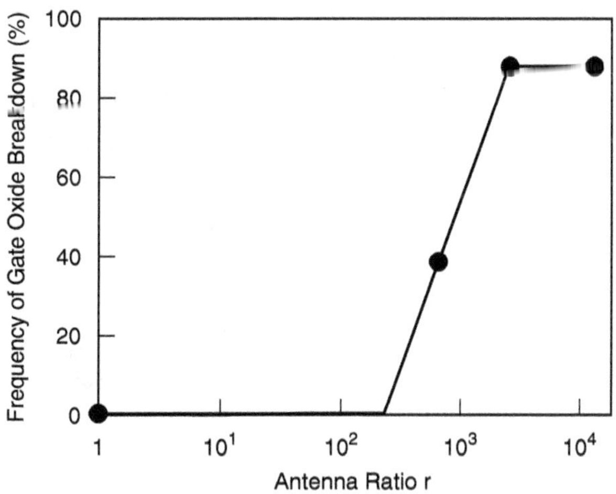

Fig. 5.9 Antenna ratio dependence of gate oxide breakdown frequency for the wafer treated with the barrel asher [10]

the self-bias voltage (V_{dc}) is larger at the wafer peripheral than at the center. As a result, as Fig. 5.10 shows, the gate voltage (V_g) distribution becomes nonuniform across the wafer. On the other hand, the Si substrate has a constant potential (V_s) across the wafer. Therefore, a potential difference is induced between the gate electrodes and the Si substrate. This potential difference ($V_g - V_s$) corresponds to the wafer charging monitored with an MNOS capacitor as the flatband voltage shift (ΔV_{FB}) [10]. The potential distribution model in Fig. 5.10 shows a positive charging at the center of the wafer and negative charging at the periphery. This figure clearly explains that the potential and polarity measured by the MNOS capacitors are relatively defined with respect to the Si substrate. Figure 5.11 summarizes the mechanism of wafer charging. There are many factors that cause a nonuniform

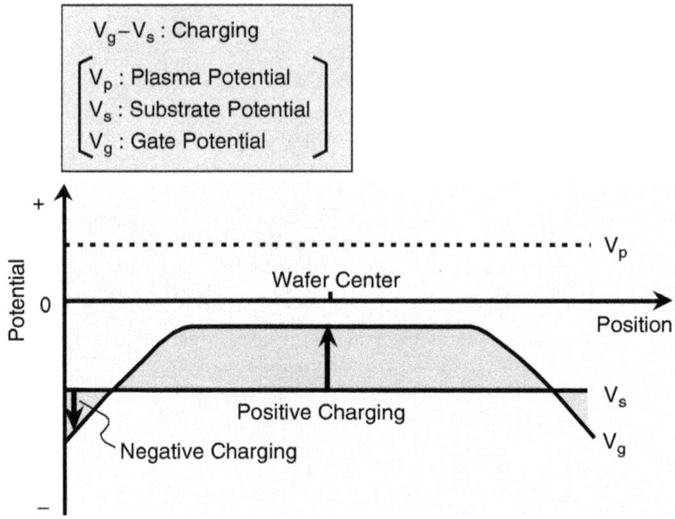

Fig. 5.10 Schematic potential distribution across the wafer when the wafer charging occurs [10]

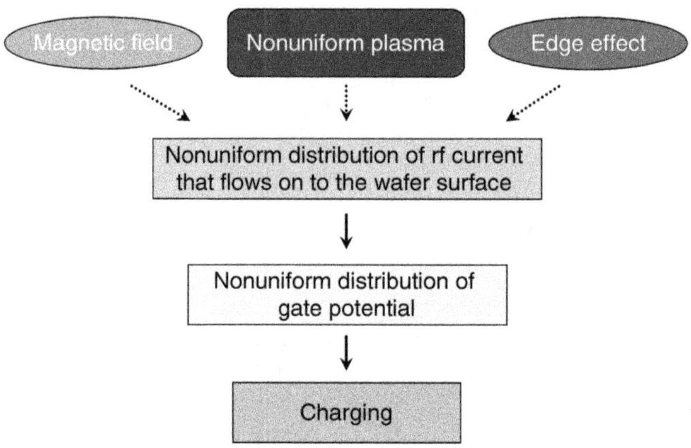

Fig. 5.11 Mechanism of wafer charging

distribution of RF current. They may include a nonuniformity in the plasma density, orientations of the electrical field and magnetic field with respect to the wafer, and their distributions. In the case of the magnetic field, the current tends to flow less in a direction orthogonal to the magnetic field, and this causes the nonuniform distribution of RF current. The next section covers the evaluation and reduction of charging damage for representative types of etching equipment. The magnitude and polarity of wafer charging were evaluated using the MNOS capacitor shown in Fig. 5.7, and the gate oxide breakdown was evaluated using the MOS capacitor shown in Fig. 5.8. The antenna ratio, gate area, and gate oxide thickness are as shown in Fig. 5.8.

5.2.3 Evaluation and Reduction of Charging Damage for Various Types of Etchers

Magnetron Reactive-Ion Etching

In a magnetron RIE, the plasma density distribution can easily become uneven because of the electrons drifting in the magnetic field and the electric field, which are orthogonal to each other. For this reason, the magnetron RIE has a disadvantage in terms of the gate oxide breakdown. Figure 5.12 shows the cross-sectional diagram of a magnetron RIE apparatus for oxide etching. A permanent magnet is set behind the anode. Figure 5.13 shows the across-the-wafer distributions of ΔV_{FB} measured with MNOS capacitors and the gate breakdown voltage measured with antenna MOS capacitors when the rotation of the magnet is halted [10]. In the absence of magnetic field, ΔV_{FB} is less than +1 V, and all the antenna MOS

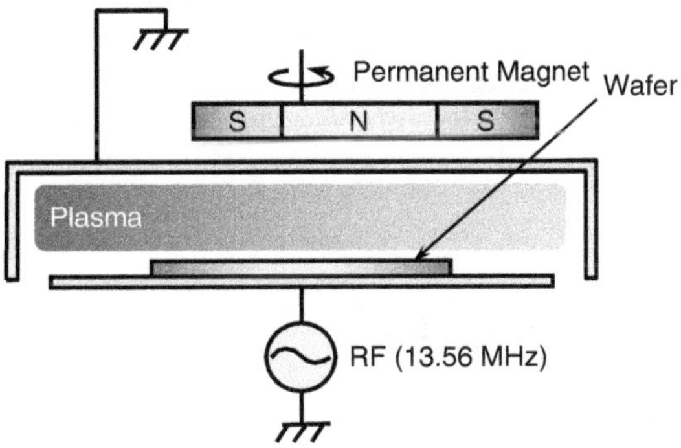

Fig. 5.12 Schematic diagram of the magnetron RIE apparatus [10]

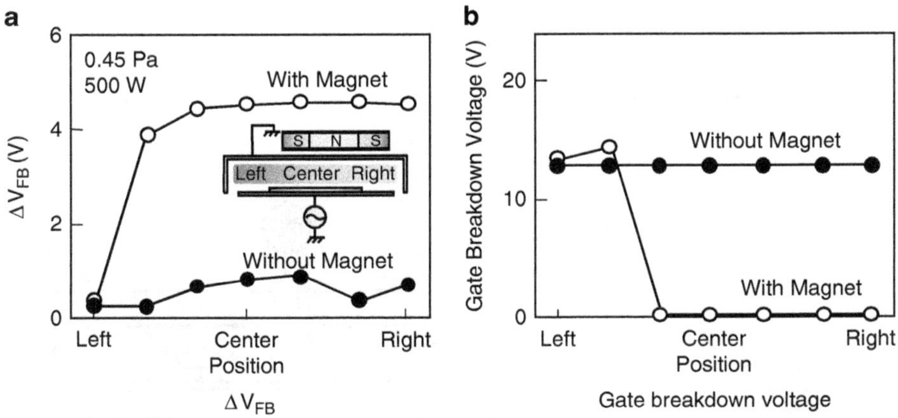

Fig. 5.13 Distribution of ΔV_{FB} and gate breakdown voltage after etching with the magnetron RIE [10]

capacitors show the intrinsic breakdown voltage. In other words, the gate oxide breakdown did not occur during plasma treatment. On the other hand, when the magnetic field is applied, a positive charging, corresponding to $\Delta V_{FB}=+4.5$ V, occurs except near the left edge of the wafer, and the gate oxide breakdown occurs in this area. It has been reported that the wafer charging in the magnetron RIE can be reduced by improving the uniformity of plasma with a curved magnetic field [11]. However, it becomes increasingly difficult to obtain a uniform plasma, as the wafer diameter increases. Therefore, the magnetron RIE is a type of etcher that has little margin in terms of gate oxide breakdown.

Parallel-Plate Plasma Etcher

On the other hand, in a CCP-type parallel-plate plasma etcher, it is easy to generate a uniform plasma across the wafer, and wafer charging can be kept low. Figure 5.14 shows the cross-sectional diagram of a parallel-plate plasma etcher for oxide etching, and Figure 5.15 shows the across-the-wafer distribution of ΔV_{FB} and the gate breakdown voltage [12]. As Fig. 5.15a shows, the gate electrodes are negatively charged, but the magnitude is small, around –0.3 to –1 V in ΔV_{FB}. Furthermore, as Fig. 5.15b shows, all the antenna MOS capacitors show the intrinsic breakdown voltage. In other words, the gate oxide breakdown did not occur during plasma treatment.

ECR Plasma Etcher

In an ECR plasma etcher, simply generating a uniform plasma is insufficient to prevent the wafer charging. Figure 5.16 shows the cross-sectional diagram of an ECR plasma etcher, and Fig. 5.17 shows the across-the-wafer distribution of ΔV_{FB} after plasma treatment in this etcher [10]. When no RF power is supplied to the wafer stage, ΔV_{FB} is zero, and there is no wafer charging. This indicates that a uniform plasma has been generated. However, when a 13.56-MHz RF power is supplied, a positive charging occurs at the center of the wafer. Electron currents

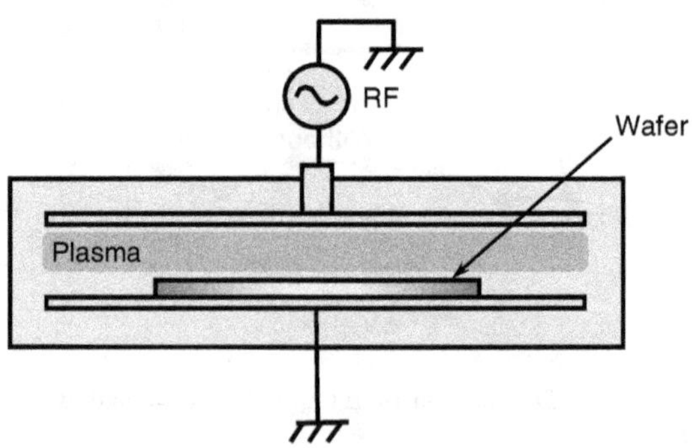

Fig. 5.14 Schematic diagram of the parallel-plate plasma etcher [12]

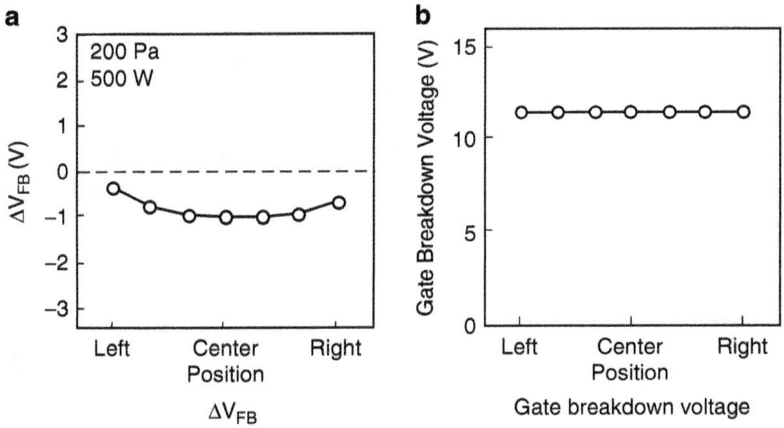

Fig. 5.15 Distribution of ΔV_{FB} and gate breakdown voltage after etching with the parallel-plate plasma etcher [12]

Fig. 5.16 Schematic diagram of the ECR plasma etching apparatus [10]

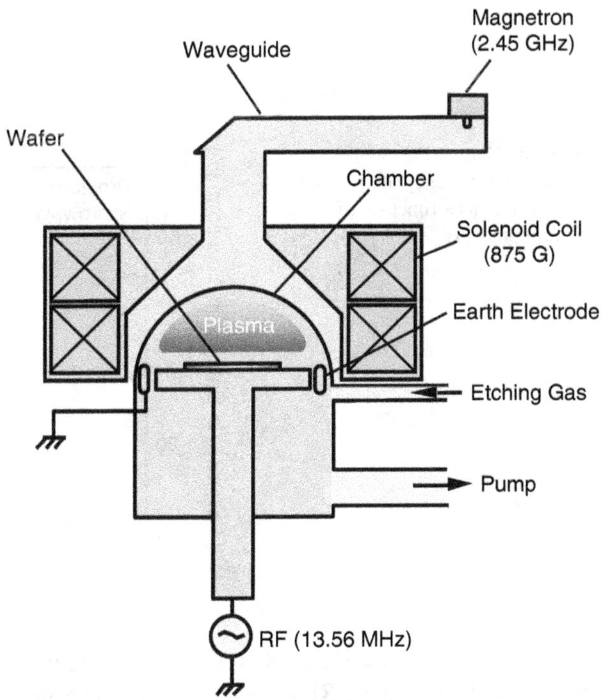

flow from the ground electrode onto the wafer surface near the periphery of the wafer, while fewer electron currents flow onto the wafer surface around the center of the wafer due to the high plasma resistivity normal to the magnetic fields. Therefore, V_{dc} becomes larger at the periphery of the wafer and smaller at the center of the wafer. Then V_g becomes nonuniform across the wafer and wafer charging occurs [10].

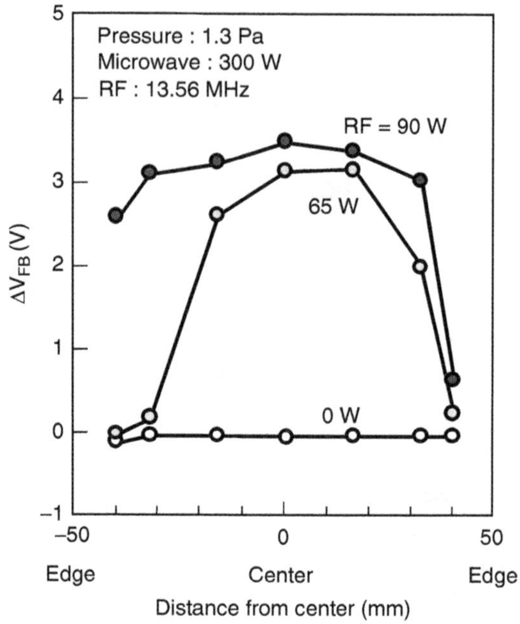

Fig. 5.17 Distribution of ΔV_{FB} after etching with the ECR plasma etcher [10]

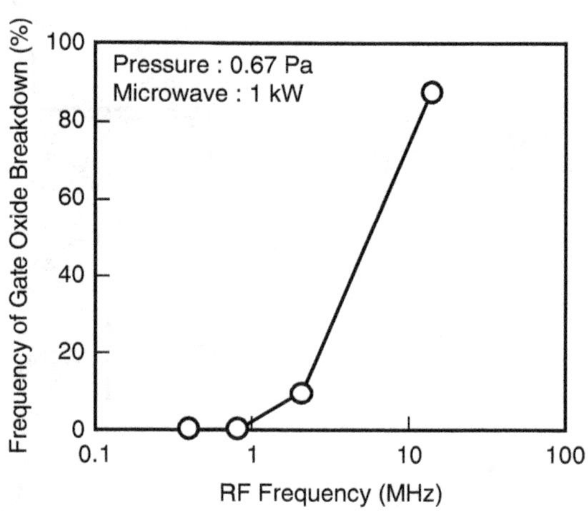

Fig. 5.18 Gate oxide breakdown frequency after ECR etching as a function of RF frequency [10]

The author has discovered that wafer charging in the ECR plasma etcher can be reduced by lowering the RF frequency [10]. The results are shown in Fig. 5.18 [10]. The gate oxide breakdown frequency decreases as the RF frequency is lowered, and no gate oxide breakdown is observed below an RF frequency of 800 kHz. The reduction in wafer charging is thought to be due to a larger ion sheath impedance at the lower frequency. This, in turn, makes the resistance due to the magnetic field negligible, and the distribution of the RF current flowing onto the wafer surface becomes more uniform. Figure 5.19 shows the maps of the gate oxide failures for

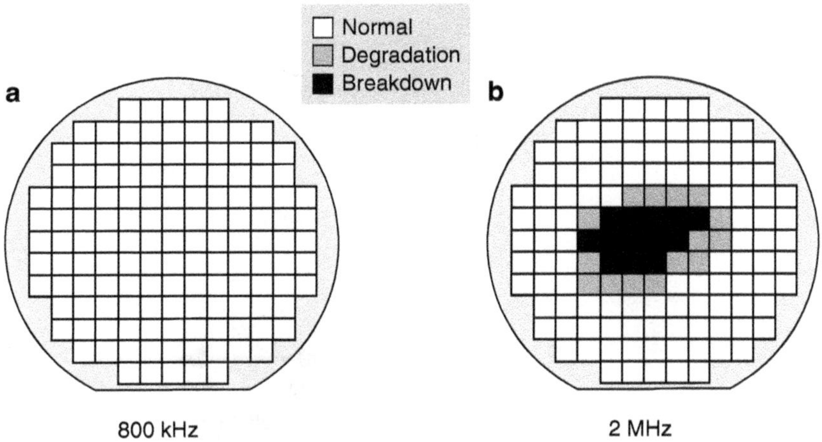

Fig. 5.19 Wafer maps of the gate oxide breakdown and degradation after ECR plasma etching [3]

Fig. 5.20 Schematic diagram of TCP® plasma etching apparatus [13]

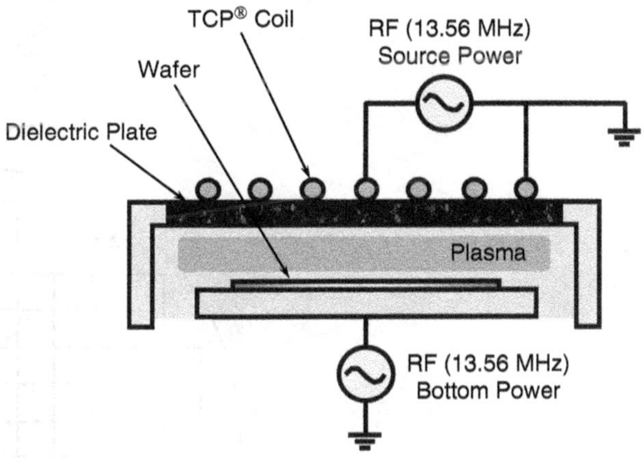

200-mm wafers etched at 800 kHz and 2 MHz [3]. At 2 MHz, the gate oxide breakdown occurs at the center of the wafer, and the area surrounding the center shows the degradation of the gate oxide. On the other hand, no gate oxide breakdown or degradation is observed across the entire wafer at 800 kHz.

Inductively Coupled Plasma Etcher

Figure 5.20 shows the cross-sectional diagram of a transformer-coupled plasma (TCP®) etcher, and Fig. 5.21 shows the bottom power dependence of the gate oxide breakdown frequency [13]. The RF frequency is 13.56 MHz for both the power supply for plasma generation (source power) and the power supply for ion energy

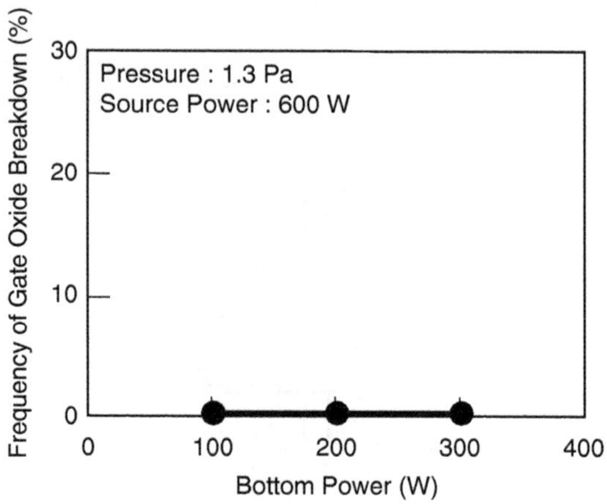

Fig. 5.21 Gate oxide breakdown frequency after TCP® plasma etching as a function of bottom power [13]

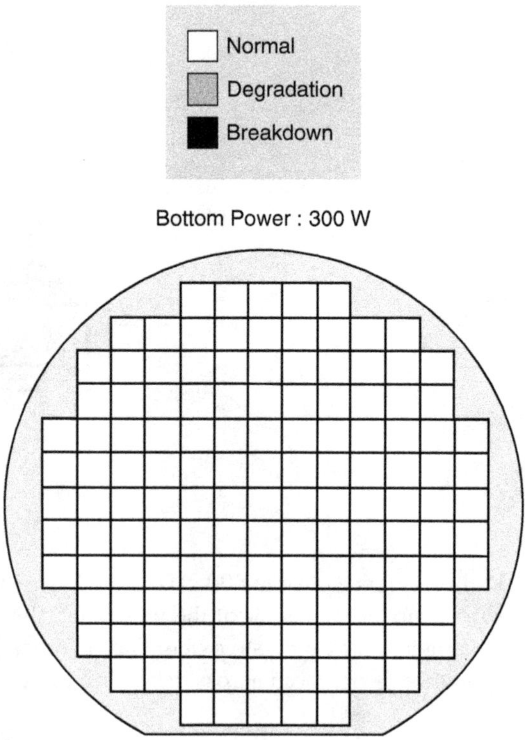

Fig. 5.22 Wafer map of the gate oxide breakdown and degradation after TCP® plasma etching [13]

control (bottom power). Figure 5.21 shows that no gate oxide breakdown occurs up to the bottom power level of 300 W. Figure 5.22 shows the map of the gate oxide failure on a 200-mm wafer, when the bottom power is 300 W [13]. No degradation or breakdown of the gate oxide is observed across the entire 200-mm wafer. In the TCP® plasma etcher, no gate oxide breakdown occurred even though the RF frequency of the bottom power is 13.56 MHz. This indicates that the magnetic field on the wafer is negligible in the TCP® plasma etcher.

As discussed so far, the wafer charging can be prevented by making the distribution of the RF current flowing on the wafer surface uniform. By doing so, it is possible to eliminate the wafer charging even for a high-density plasma using a strong magnetic field as in the ECR etcher.

5.2.4 Pattern-Dependent Gate Oxide Breakdown

As devices are scaled, the gate oxide breakdown due to the pattern-related charging becomes more serious. This is called *electron shading damage*, and the gate oxide breakdown becomes more likely to occur with denser patterns [14]. Figure 5.23 shows a model [15]. While incoming ions travel in the vertical direction, incoming electrons travel at an angle and cause the sidewalls of the resist to be charged negatively. Because electrons are then repelled by the negative charges and are no longer able to reach the bottom of the patterns, there will be an excessive amount of ion currents. As a result, the metal lines are charged positively and cause an injection of excessive currents, which in turn damages the gate oxide film. This effect becomes more significant with narrower pattern spaces. This is why the gate oxide breakdown is more likely to occur with dense patterns.

An effective way to reduce the electron shading damage is to reduce the electron temperature. This can be done by widening the gap between the electrodes and/or

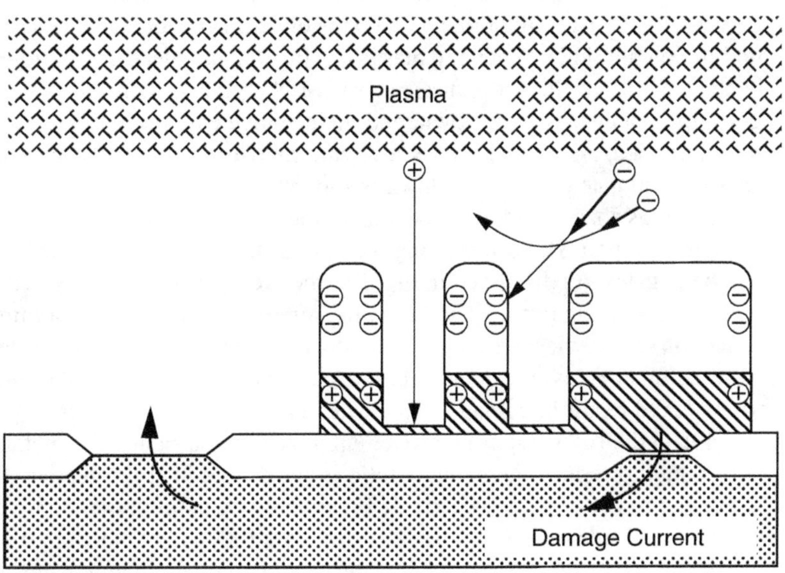

Fig. 5.23 Model for electron shading damage [15]

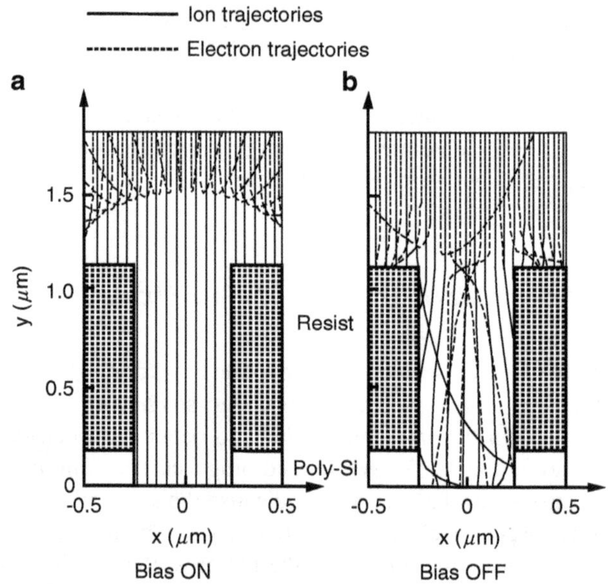

Fig. 5.24 Ion and electron trajectories in time modulation bias method [18]

raising the pressure [16]. More effective methods to reduce the electron shading damage have been proposed. One is based on the time modulation of the plasma discharge [17], and the other one is based on the time modulation of the RF bias [18]. The former is aimed at neutralizing the positive charging with negative ions generated at the discharge-off period [17]. With the latter, the RF bias is turned on and off periodically at several kilohertz; it is called the *time modulation bias method* [18]. Figure 5.24 shows a simulation result on the electron and ion trajectories under a time modulation bias [18]. When the bias is on (Fig. 5.24a), electrons are repelled, and only the ions travel in and cause a positive charging. When the bias is off (Fig. 5.24b), the ion energy goes down, and the negative charges at the resist sidewalls are neutralized. As a result, no stress current will flow into the gate at the bias-off period, and gate oxide breakdown is suppressed.

The most serious charging damage occurs during the etching and ashing steps for the metal lines of logic LSI, because long metal lines that could act as antennae are connected to the gates. As discussed in Fig. 5.9, the breakdown frequency increases with an increased antenna ratio. To address this, some of the measures that might be implemented on the device structures include the antenna rules and protective diodes. The antenna rules are created based on experimenting with the antenna ratios, which could result in damage to the gate, and the metal lines with segments that could violate the rules might be broken up to satisfy the rules. The metal lines might be rerouted to reduce the antenna ratio in the design [19]. Protective diodes might be created to prevent the gate oxide breakdown by making the damage currents flow into the Si substrate through a diffusion layer [20]. All of these approaches, however, end up imposing restrictions on the layout rules and cannot always be used. They should be combined with effective measures against the wafer charging through appropriate etching hardware and process design.

References

1. G. S. Oehrlein, Proc. Symp. Dry Process, p.59 (1986).
2. G. S. Oehrlein, R. M. Tromp, Y. H. Lee, and E. J. Petrillo: Appl. Phys. Lett., 45 (4), 420 (1984).
3. K. Nojiri, T. Mizutani: Oyo Buturi, Vol. 64, No.11, p.115 (1995).
4. K. Nojiri and E. Iguchi: J. Vac. Sci. & Technol., B 13, 1451 (1995).
5. K. Hashimi, D. Matsunaga, M. Kanazawa: Ext. Abstr. 40th Spring Meeting of The Japan Society of Applied Physics, No. 2, p.616 (1993).
6. K. Hashimi, D. Matsunaga, M. Kanazawa: Ext. Abstr. 54th Fall Meeting of The Japan Society of Applied Physics, No. 2, p.536 (1993).
7. N. Aoto, M. Nakamori, H. Hada, T. Kunio, and E. Ikawa: Ext. Abstr. Int. Conf. Solid State Devices and Materials, p.101 (1993).
8. K. Tsunokuni, K. Nojiri, S. Kuboshima and K. Hirobe: Ext. Abstr. 19th Conf. Solid State Devices and Materials, p.195 (1987).
9. Y. Kawamoto: Ext. Abstr. 17th Conf. Solid State Devices and Materials, p.333 (1985).
10. K. Nojiri and K. Tsunokuni: J. Vac Sci. Technol., B 11, 1819 (1993).
11. S. Nakagawa, T. Sasaki, and M. Mori: Proc. Symp. Dry Process, p.23 (1993).
12. K. Nojiri: Semiconductor World, p.86, October (1992)
13. K. Nojiri: Charging Damage in Semiconductor Process, p.51 (REALIZE INC, 1996).
14. K. Hashimoto: Jpn. J. Appl. Phys., 32, 6109 (1993).
15. K. Hashimoto: Dig. Pap. Int. MicroProcess Conf., p.146 (1995).
16. K. Nojiri,K. Kato and H. Kawakami, Proc. 4th Int. Symp. Plasma-Process Induced Damage, p.29 (1999).
17. H. Ohtake, S. Samukawa, K. Noguchi and T Horiuchi: Proc. Symp. Dry Process, p.97 (1998).
18. K. Nojiri, N. Mise, M. Yoshigai, Y. Nishimori, H. Kawakami, T. Umezawa, T. Tokunaga and T. Usui: Proc. Symp. Dry Process, p.93 (1999).
19. K. Noguchi: Electronic Journal 8th Technical Symposium, p.31 (1997).
20. M. Takebuchi, K. Yamada, T. Nishimura, K. Isobe, T. Uemura, T. Fujimoto, M. Arakawa, S. Mori, A. Kimitsuka and K. Yoshida: Tech. Dig. Int. Electron Devices Meet., p.185 (1996).

Chapter 6
Latest Dry Etching Technologies

This chapter reviews the latest etching technologies, including Cu damascene etching, low-κ etching, metal gate/high-κ etching, and FinFET etching. There are also discussions on the various methods of Cu damascene etching as well as the methods of preventing damage to low-κ films, which become a serious issue at the 32-nm node and beyond. Furthermore, the discussion includes a review of the double-patterning technology, which is the hot topic of the moment, and three-dimensional integrated circuit (3D IC) etching for the 16-nm node and beyond.

6.1 Copper Damascene Etching

The speed of logic LSI devices has been increased mainly via scaling of transistors. However, metal lines became the limiting factor on the speed of LSI devices starting at the 0.25-μm node [1]. Figure 6.1 shows the reason for this. With more scaling (and higher levels of integration), the cross section of the metal line becomes smaller, while the metal lines become longer. As a result, the metal line resistance R goes up. At the same time, the distance between the metal lines becomes shorter, and the metal-to-metal capacitance C goes up. Because the delay time in the metal lines is proportional to RC, the delays go up with scaling (higher device density) and slow down the speed of the LSI devices. As a solution, Cu, which has a lower resistivity, was introduced as an alternative to the conventional Al in order to reduce the metal line resistance. While the resistivity of Al is approximately 2.7 μΩcm, the resistivity of Cu is lower, at approximately 1.7 μΩcm. Furthermore, an insulating film with a lower dielectric constant (low-κ film) was introduced as an alternative to the conventional oxide film for reducing the inter-metal capacitance. While the oxide film has a dielectric constant κ of 4.1, the κ-value of the low-κ film is 3 or less. A new technology called damascene was introduced for forming the Cu metal lines.

K. Nojiri, *Dry Etching Technology for Semiconductors*,
DOI 10.1007/978-3-319-10295-5_6

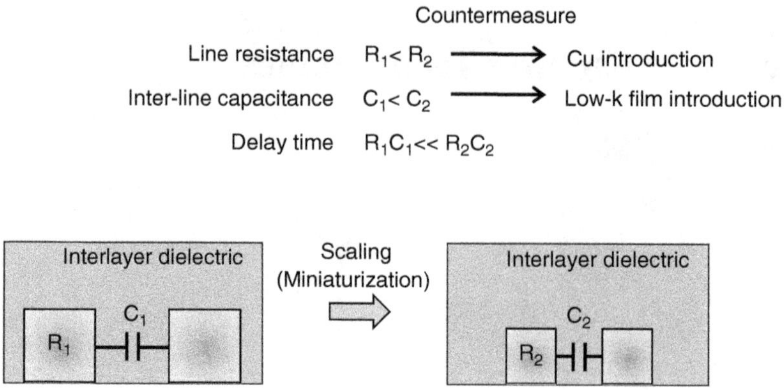

Fig. 6.1 Delay time in metal line increases as the dimension shrinks

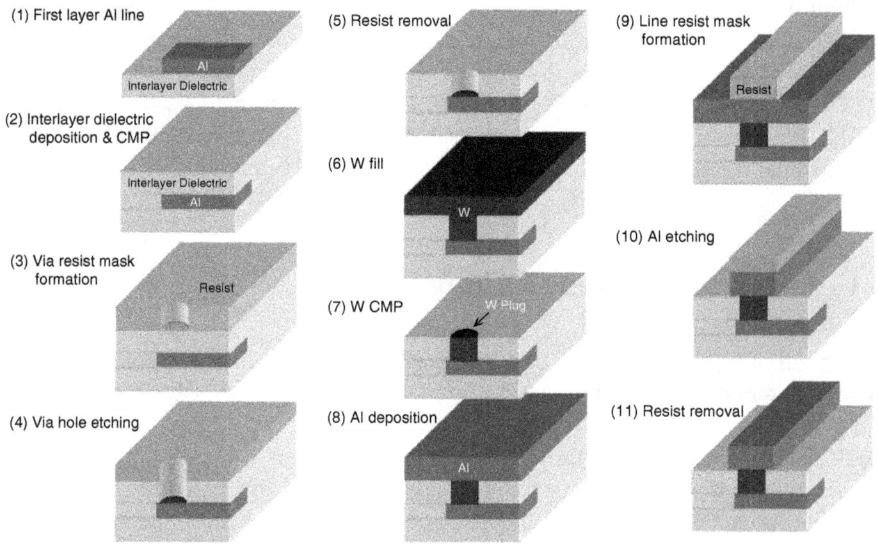

Fig. 6.2 Process flow for Al multilevel interconnects

As previously mentioned in Sect. 3.3.3 in Chap. 3, the new technology was needed because it is extremely difficult to form Cu metal lines by etching. We will describe the multilevel interconnect process with damascene technology through comparisons with a conventional Al multilevel interconnect process.

Figure 6.2 shows the process flow for Al multilevel interconnects. The steps are as follows:

1. The first-layer Al metal lines have been formed.
2. An interlayer insulating film is deposited and planarized by chemical mechanical polishing (CMP).

3. A resist mask for the via holes is formed using the lithography technology.
4. Then via holes are opened by an oxide film etcher.
5. The resist is stripped off by an asher.
6. Tungsten is filled into the via holes by CVD.
7. Next, W at the flat parts is removed by polishing with the CMP process, and W plugs are formed.
8. The second-layer Al is deposited by sputtering.
9. A resist mask for the metal lines is formed.
10. The metal lines are etched with an Al etcher.
11. The resist is removed with an asher. The second-layer Al metal lines are thus formed.

Next, the damascene process is illustrated in Fig. 6.3. The damascene process refers to a technology in which Cu is deposited into trenches dug in an insulating film and polished by CMP in order to form the metal lines. The damascene process could be single-damascene, with the metal lines and via holes formed separately, or dual-damascene, with the metal line trenches and via holes filled with Cu at the same time. The dual-damascene technology requires fewer process steps than the single-damascene process and can give a lower-cost process. There are several

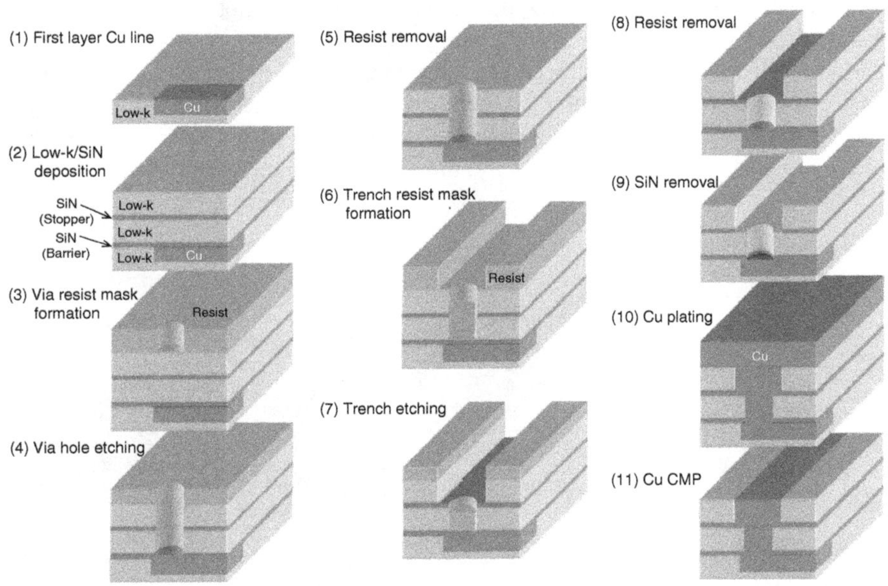

Fig. 6.3 Process flow for low-κ Cu dual damascene (via first)

schemes of dual-damascene, and Fig. 6.3 shows the via-first scheme that has become the mainstream technology these days, whose steps follow:

1. The first-layer Cu metal lines have been formed.
2. A barrier film (SiN), for preventing the diffusion of Cu, is deposited on top of Cu, and a low dielectric constant film (low-κ film), in which the via holes are to be formed, is then deposited on top. Next, an SiN film, which functions as an etch stopper, and a low-κ film, in which the trenches are to be formed, are deposited one after the other.
3. A resist mask for the via holes is formed with the lithography technology.
4. Then via holes are opened by an oxide etcher.
5. The resist is stripped off with an asher.
6. A resist mask for the trenches is then formed.
7. Trenches are then etched by an oxide etcher.
8. The resist is stripped off to yield a geometric shape in which the trenches for the metal lines and the via holes connect with one another.
9. Next, SiN is removed to expose Cu (first-layer Cu) at the bottom of the via holes.
10. Copper is filled into the via holes and trenches by plating.
11. Cu at the flat parts is removed by CMP, and the second-layer Cu metal lines are formed.

When the SiN film is used as a stopper layer for trench etching, the effective dielectric constant goes up, and the benefit of using a low-κ film is reduced. For this reason, the stopper layer for etching is not used in most cases these days. In that case, it is necessary to achieve a stable etch rate, because the trench depth is determined by the etching depth. Figure 6.4 shows an example of the dual-damascene etching. The example here is SiO_2 damascene etching and not a low-κ damascene etching. The stopper SiN for trench etching is not used here.

As thus described, it is not necessary to etch Cu to form the Cu metal lines with the damascene process. Therefore, a metal etcher is no longer needed, and an oxide etcher is used instead. In other words, the introduction of the damascene process resulted in less use of metal etchers and more use of oxide etchers.

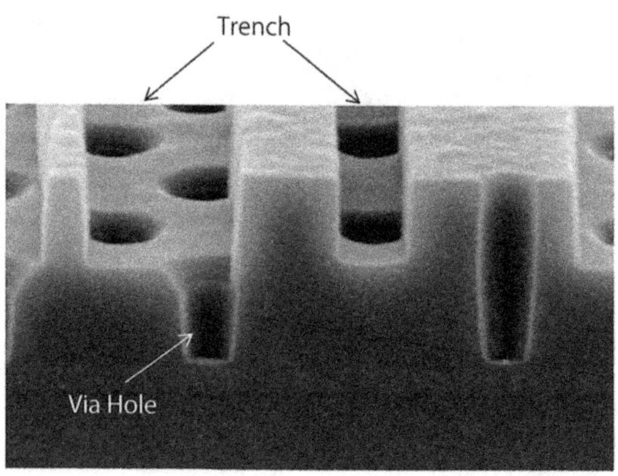

Fig. 6.4 Example of dual-damascene etching

6.2 Low-κ Etching

The Cu dual-damascene technology, in combination with SiO_2 film, was first introduced at the 0.18-μm technology node. With the 0.13-μm technology node, fluorine-doped SiO_2 (FSG) with a dielectric constant κ of 3.6 was introduced. Fluorine-doped SiO_2 can be etched with almost the identical etching conditions as SiO_2. The low-κ film, with κ=3.0 or less, was introduced in earnest at the 90-nm technology. A process combination involving low-κ film makes the overall process technology more challenging, because the etching of low-κ film itself is a difficult process.

The low-κ film introduced for the 90-nm logic LSI had a κ-value of around 2.9. The deposition methods used for the low-κ films have included spin-on glass (SOG) and chemical vapor deposition (CVD). Both result in films containing the methyl group CH_3, which reduces the κ-value. For the etching and ashing of low-κ films, special attention must be paid to the side effects of oxygen. Figure 6.5 illustrates the reason for this requirement; it shows the reaction between the low-κ film and the oxygen atom. This is a simulation result using the molecular orbital method [2].

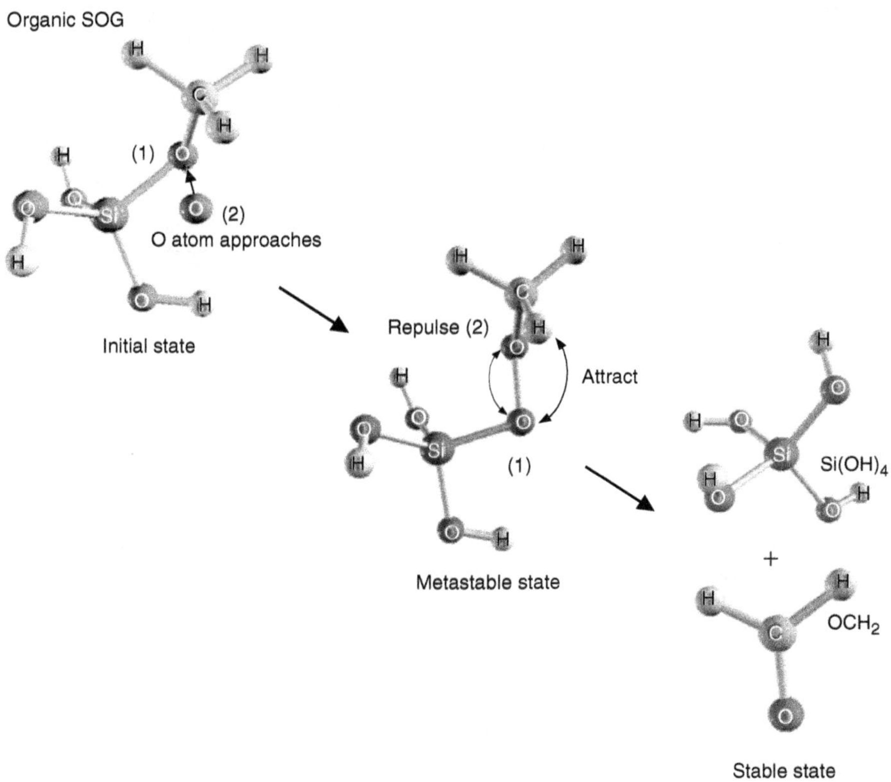

Fig. 6.5 Reaction between methylsiloxane-based organic SOG and oxygen atom [2]

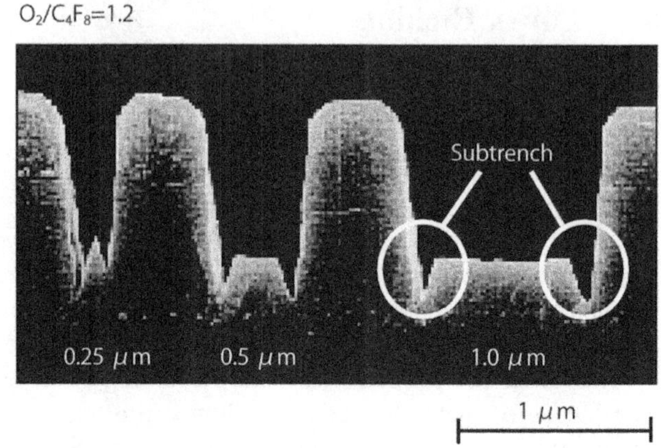

Fig. 6.6 Subtrench formation during low-κ film (methylsiloxane-based organic SOG) etching with C_4F_8/O_2/Ar gas chemistry [2]

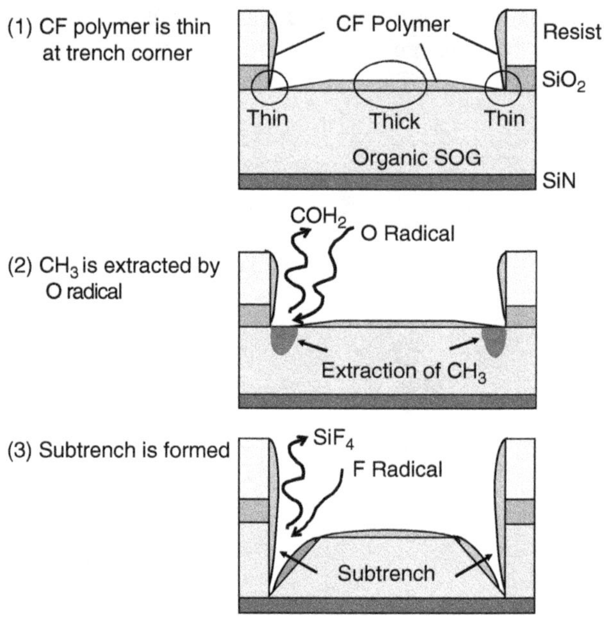

Fig. 6.7 Mechanism of subtrench generation during low-κ film (methylsiloxane-based organic SOG) etching with C_4F_8/O_2/Ar gas chemistry [2]

The low-κ film here is a methylsiloxan-based organic SOG containing CH_3 in it, and the figure shows what happens when an oxygen atom approaches. The approaching oxygen atom extracts the CH_3. As a result, the κ-value goes up, and the etch rate of the areas that became porous increases. During trench etching, oxygen in the etching gas can cause subtrenches. Figure 6.6 shows an example of the subtrenches [2]. Here a methylsiloxan-based organic SOG is etched with a C_4F_8/O_2/Ar gas chemistry. Figure 6.7 illustrates the mechanism of how the subtrenches are created [2]. As Fig. 6.7 (1) shows, the thickness of CF polymers deposited inside the trench is not uniform. It is thinner at the bottom corners of the trench. As a result, these areas tend to be etched easily by the oxygen radicals, and the underlying organic SOG becomes exposed to the oxygen radicals. The etch rates at these areas are

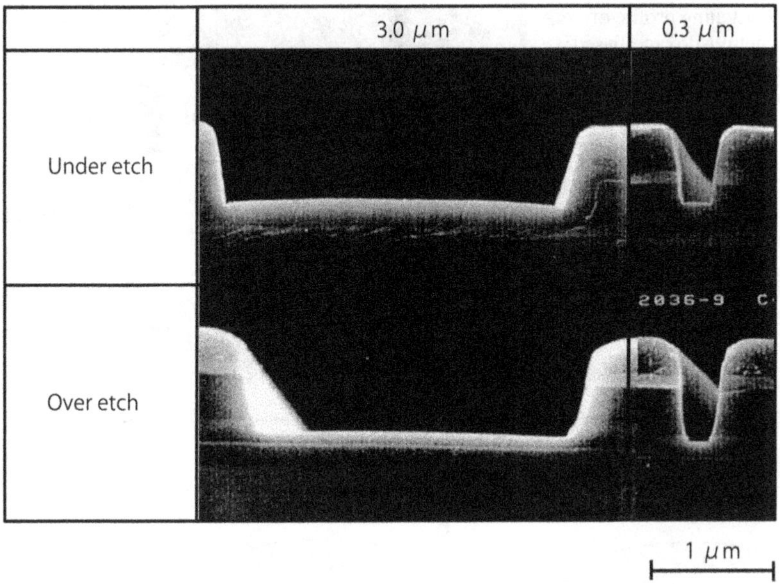

Fig. 6.8 Cross-sectional profile of low-κ (methylsiloxane-based organic SOG) trench etched with $C_4F_8/O_2/Ar$ gas chemistry [2]

thus enhanced through the process described above, and the subtrenches are then generated. The formation of subtrenches can be prevented by optimizing the O_2 concentration [3], or using N_2 instead of O_2 [2]. Figure 6.8 shows the results of methylsiloxan-based organic SOG etching with $C_4F_8/N_2/Ar$ gas chemistry. No subtrenches are observed during both the underetching and overetching periods. Figure 6.9 shows an example of dual-damascene etching on the same organic SOG using the optimized concentration of O_2 and C_4F_8 [3].

The low-κ film containing CH_3 requires special attention with ashing as well. As Fig. 6.10a shows, the oxygen radicals extract the CH_3 during ashing with an oxygen plasma [4]. This phenomenon is referred to as low-κ damage. The damaged portion will easily be etched off with light etching in dilute hydrofluoric acid. When a low-κ film is damaged, its κ-value goes up. For this reason, ashing after trench etching requires special care, compared with ashing after via hole etching. Ashing after trench etching is more critical because the capacitance between the trenches has a direct impact on the delay time in metal lines. As shown in Fig. 6.10b, this problem can be resolved by forming dense SiO_2 layers on the sidewalls using oxygen ions [4]. Specifically, it is effective to increase the ion concentration by lowering the pressure down to around 0.133 Pa (1 mTorr) during ashing, or to use a bias ashing that attracts the oxygen ions by applying a bias to the wafer stage.

Another type of low-κ film is an organic polymer film. It is generally etched with an O_2/N_2 gas using a hard mask such as SiO_2 [5], but this tends to lead to a bowing profile during overetching. The N_2/H_2 or NH_3 gas is more effective for higher-precision processing [6]. The resist stripping process can be a problem also for the

Fig. 6.9 Etching profile of low-κ (methylsiloxane-based organic SOG) dual damascene [3]

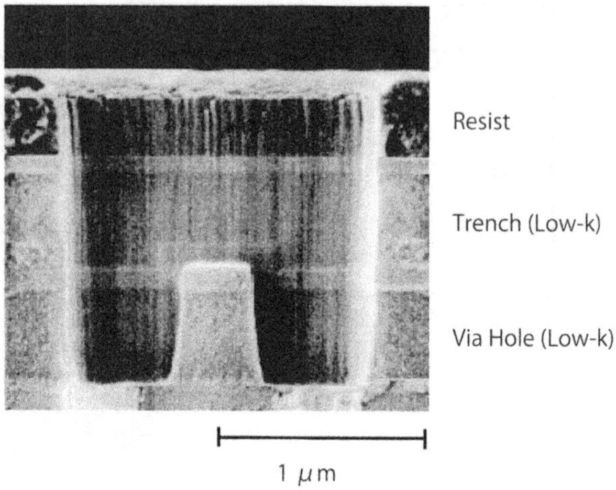

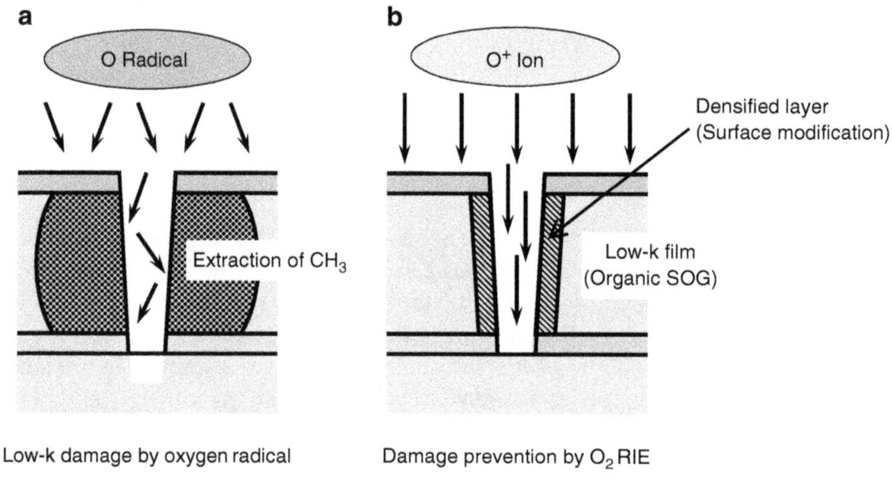

Fig. 6.10 Prevention of low-κ damage during ashing [4]

organic polymer films. There has been a report on the process in which the hard mask is opened using a resist mask, and then the resist is etched off during the etching of organic film by O_2 reactive-ion etching (RIE) [7].

For the actual multilevel interconnects, the dual-damascene process described earlier is repeated to create a multitude of layers. Figure 6.11 shows an example with 12 layers. Because the metal line pitches are narrower with the local metal lines and intermediate metal lines, the low-κ films with the lowest κ-value are used on these layers. The low-κ film with a slightly higher κ-value is used on the semi-global metal lines, and the regular SiO_2 film is used on the global metal lines. In this example, the 12th layer is the top layer metal line. This layer includes bonding pads, and the regular Al process is used. In this case, the metal lines are formed by the conventional Al etching technology, instead of a damascene process.

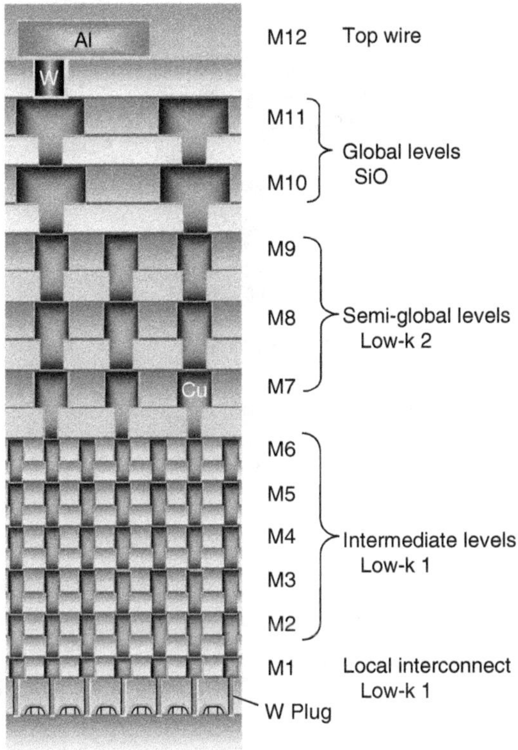

Fig. 6.11 Multilevel interconnects using Cu dual damascene

6.3 Porous Low-κ Damascene Process

Starting with the 45-nm technology node, porous low-κ films with κ-values of 2.5 or less were introduced. Damage induced by ashing became more serious with the porous low-κ film, and the solutions described earlier were no longer adequate. For this reason, it was necessary to change the dual-damascene flow itself. Figure 6.12 shows an example. The key aspect of this example is a metal hard mask used for trench etching. TiN is one of the widely used materials. A process flow is as follows:

1. The resist patterns for trenches are formed on TiN first.
2. The metal hard mask (metal HM) is etched first. A metal etcher is used here to etch TiN with Cl-based gas chemistry.
3. Then the resist is stripped off with an asher. Because the resist is stripped off at this step, there is no step for removing the resist after trench etching, and damage, therefore, is prevented.
4. Next, the resist patterns for the via holes are formed.
5. "BARC" in the figure refers to the bottom antireflection coating. Via holes are then etched
6. The resist is removed.
7. The trenches are etched using the metal hard mask formed earlier.
8. Then the barrier film is etched off, and the dual-damascene etching process is completed.

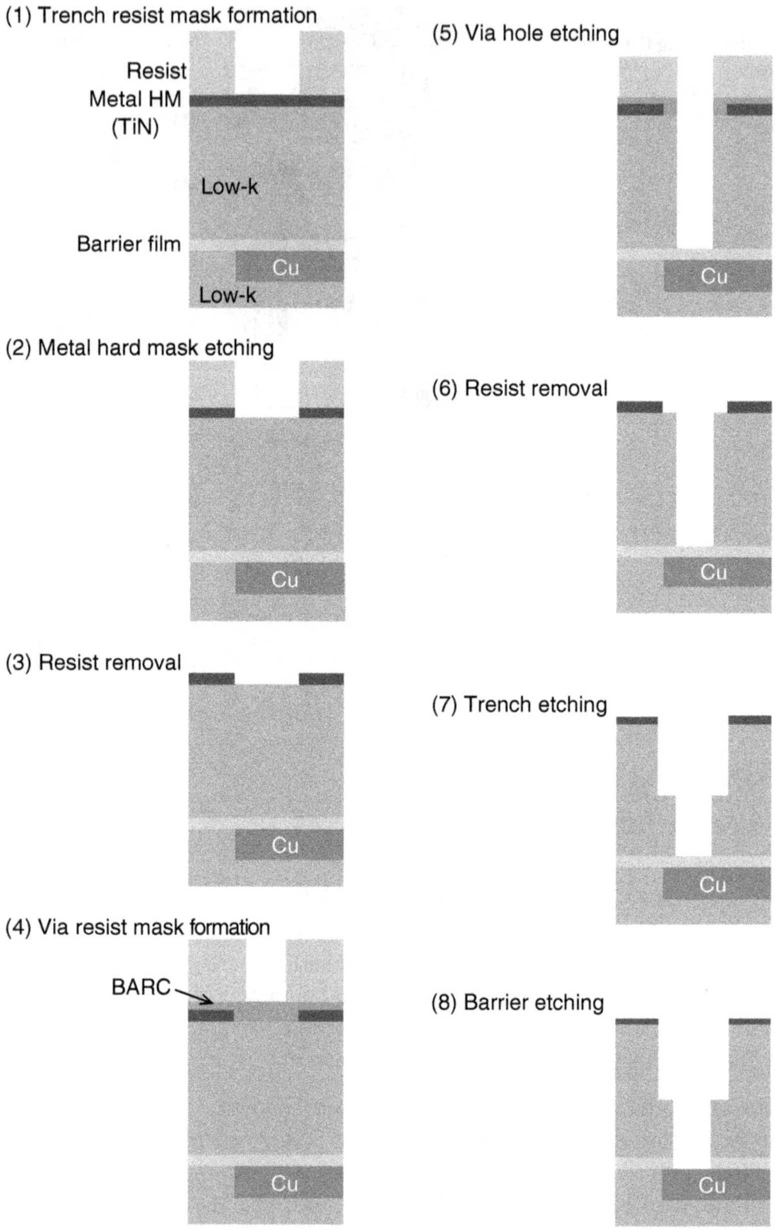

Fig. 6.12 Dual-damascene process flow using metal hard mask

The reason to use a metal hard mask like TiN for trench etching in this process is to achieve a large selectivity to the porous low-κ film. This process is sometimes referred to as the trench-first scheme, because the trench mask is formed before the via holes.

6.4 Metal Gate/High-κ Etching

As devices are scaled, the gate insulating film becomes thinner. Specifically, the equivalent oxide thickness (EOT) goes down to around 1.0 nm at the 45-nm node. An increase in leakage current is an issue here. While SiON worked well enough through the 65-nm node, the gate leakage current became no longer negligible at the 45-nm node and beyond. As a solution, the high-κ film was introduced as an alternative to the SiON film at the 45-nm node. When the high-κ film is used, the higher dielectric constant helps achieve the same EOT with thicker film. For this reason, the increase in leakage current can be suppressed. A variety of films were investigated as high-κ film candidates. The choices have converged on the Hf-based films, such as HfO_x and $HfSiO_x$ for the 45- and 32-nm nodes.

There was also a problem with the gate electrode material. A depletion layer is formed in the conventional poly–Si gate and increases the effective gate insulating film thickness. For this reason, metal gates, in which no depletion layer is formed, were introduced at the 45-nm node. Various materials were also investigated, and the choices have converged on TiN and TaN for the 45- and 32-nm nodes.

The metal gate/high-κ structure was introduced for the first time at the 45-nm node [8] and came to be widely used at the 32-nm node. A metal gate/high-κ process can either be gate-first [9] or gate-last [8], as shown in Fig. 6.13. With the gate-first scheme, the process flow itself is the same as the conventional poly–Si gate process. Fig. 6.13a shows the process flow:

1. After the metal/high-κ is deposited.
2. This stacked layer film is etched to form the gates.
3. Then the steps for source drain extension (SDE) ion implanting, spacer formation, source drain (SD) ion implantation, activation annealing, and silicidation follow for completing the gates.

The gate-first process is simple and short, and the manufacturing cost tends to be lower. On the other hand, because the high-temperature treatment steps are used for activation annealing and silicidation after the metal/high-κ gate is formed, the work function goes down, and the MOS transistor threshold voltage V_{th} goes up. Furthermore, because the etching of the metal/high-κ structure is challenging, there are many issues in the gate etching process. Specifically, these issues include etch profile control, residues, and Si recess (etching of the Si substrate).

With the gate-last method, a poly–Si gate is first formed. After the heat treatment steps such as ion implantation, activation annealing, and silicidation are completed, the poly–Si gate is removed. The high-κ film and metal are deposited into

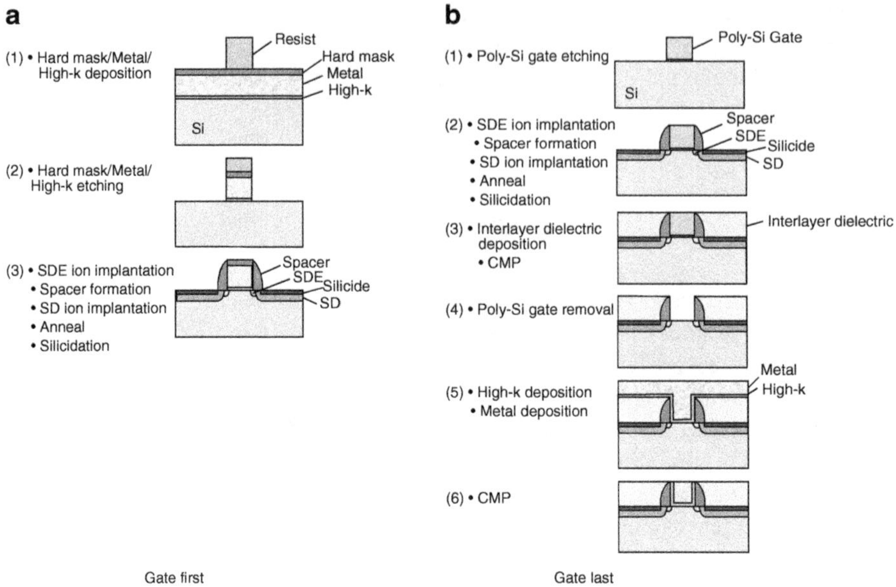

Fig. 6.13 Process flow for metal gate/high-κ

the resulting trenches and planarized by CMP, and the gates with a metal/high-κ structure are thus formed. Because the poly–Si gates are first formed and then replaced by the metal gate/high-κ, this scheme is also referred to as the replacement gate. Figure 6.13b shows the process flow:

1. First, poly–Si gates are formed by poly–Si gate etching.
2. Next, SDE ion implantation, spacer formation, SD ion implantation, activation annealing, and silicidation follow.
3. After an interlayer film is deposited, it is planarized by CMP.
4. The poly–Si gate is removed.
5. The high-κ film and metal are deposited.
6. The high-κ film and metal at the flat part are removed by CMP.

The process for forming the metal/high-κ gates is then completed. Because all the thermal steps are completed before the metal/high-κ is formed in the gate-last scheme, there is no thermal budget issue, and the process is stable. Furthermore, the poly–Si gate etching is less challenging than the etching of the metal gate/high-κ structure. On the other hand, the process is more complex, and the process flow is longer. Therefore, the manufacturing cost is higher.

This section focuses on the etching technology for the gate-first scheme, which has many challenges. Figure 6.14 shows the typical structure and materials that are used for the metal gate/high-κ [10]. The high-κ films used are hafnium oxides, including HfO_2, $HfSiO_2$, and HfSiON. The metal used is TiN or TaN, and a thin cap metal of approximately 1 nm is inserted between the metal and high-κ films to

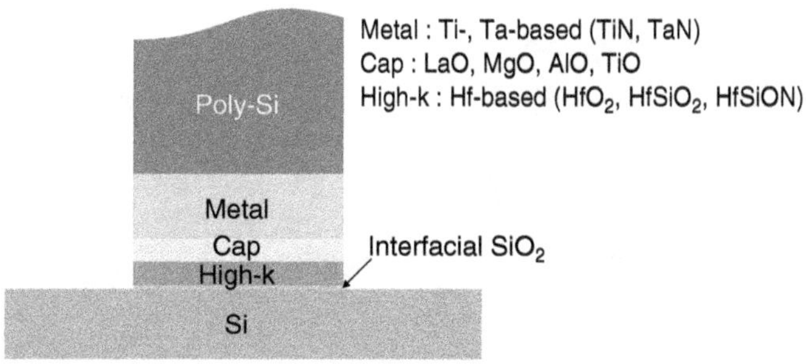

Fig. 6.14 Schematic of typical metal gate/high-κ stack [10]

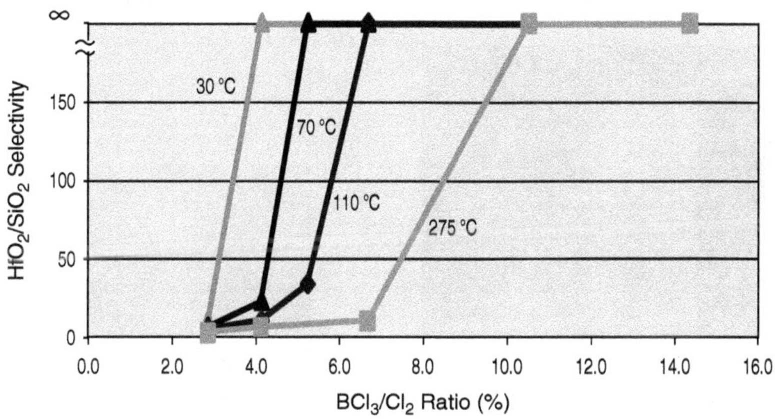

Fig. 6.15 HfO_2/SiO_2 selectivity as a function of the BCl_3/Cl_2 ratio at various temperatures [10]

adjust the MOS transistor threshold voltage V_{th}. LaO and AlO are widely used for the n-type and p-type MOS transistors, respectively. In the typical metal gate/high-κ stack, the thickness of the metal is approximately 10–20 nm, and 100-nm-thick poly–Si is placed on it. Titanium nitride and TaN can easily be etched with halogen-based gases. On the other hand, high-κ film etching is difficult because the etch byproducts are nonvolatile. For the high-κ film etching, it is necessary to keep a large selectivity to the mask and avoid damage to the underlying source and drain regions while minimizing the impacts on the gate profile. Because the hafnium chlorides have higher vapor pressures than the fluorides [11], BCl_3/Cl_2 is generally used for HfO_2 etching. With the $HfSiO_2$ high-κ film containing Si, etching is relatively easier than with HfO_2 because Si halides tend to have higher vapor pressures [11]. Figure 6.15 shows the HfO_2/SiO_2 selectivity as a function of the BCl_3/Cl_2 ratio at various temperatures. A TCP® etcher was used for this experiment [10]. When the BCl_3/Cl_2 ratio is optimized, an infinite HfO_2/SiO_2 selectivity is obtained at any temperatures between 30–275 °C. It is thought that the infinite selectivity results from the deposition of BCl_xO_y on the surface.

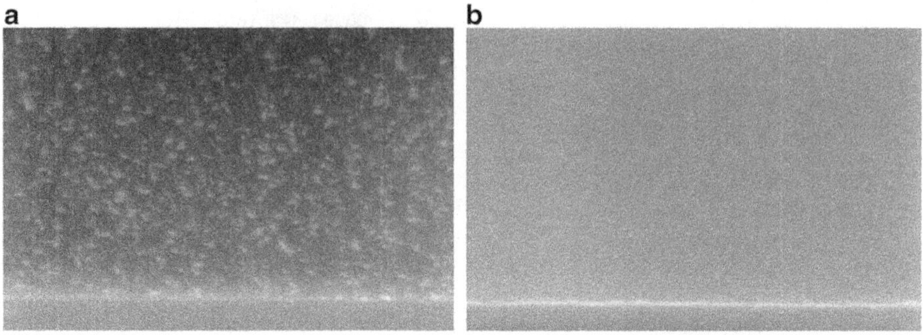

Improper cleaning shows La-based residues   Proper cleaning shows no residues

Fig. 6.16 Proper cleaning is needed for clean surface with no La-based residues [10]

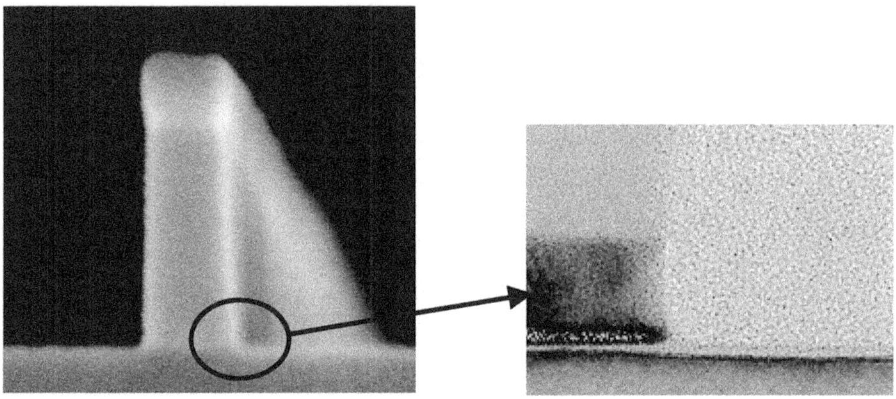

Fig. 6.17 Vertical metal gate/high-κ profile with no residue and no Si recess [10]

Another issue that needs to be addressed is the La residues. As mentioned earlier, the LaO cap on the high-κ film is used for the n-type MOS transistors. Because La etch byproducts are nonvolatile, they tend to lead to residues. These residues can usually be removed with wet cleaning after the dry etching. If the queue time between the dry etching and wet cleaning steps is long, the residues become harder to remove [12]. Therefore, the wafers should be wet cleaned immediately after dry etching. Figure 6.16 compares the results when the wet cleaning is completed appropriately and when it is not [10]. The etched profile is shown in Fig. 6.17 [10]. The gate stack consists of poly–Si/TiN. A vertical profile with no residue and no Si recesses is obtained.

6.5 FinFET Etching

A three-dimensional FinFET structure is being considered for MOS transistors for the 15-nm node and beyond [13]. As Fig. 6.18 shows, a FinFET is formed with a thin Si layer (fin) that is orthogonal to the substrate, and this layer is surrounded

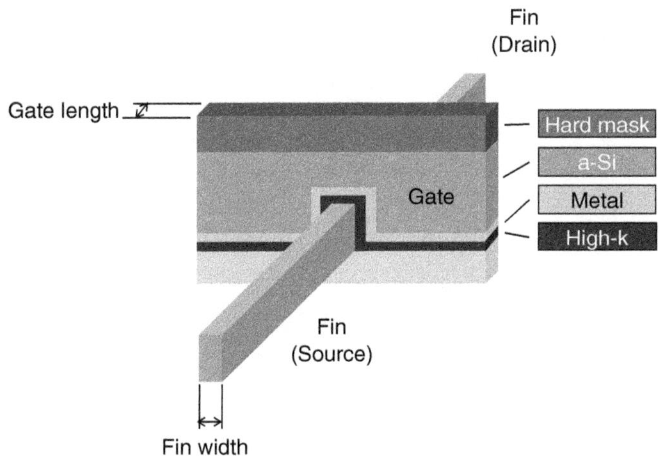

Fig. 6.18 Structure of FinFET [15]

by the gate [15]. Because of the double-gate structure, the short channel effect is suppressed, and the drain current is also large. As a result, the variability in transistor characteristics is reduced [14]. The gate length for the 15-nm node is around 20 nm, and the fin width is around 10 nm. Gate etching for the FinFET is extremely challenging, because the film tends to remain at the bottom corner of the fin.

Figure 6.19 shows a process flow for metal gate/high-κ etching for FinFETs [15]:

1. This is the cross section before poly–Si etching.
2. For the main etching, a highly anisotropic process is used for etching an 80-nm-thick poly–Si film by approximately 60 nm, but the etching is stopped before TiN is exposed. The amount of etching is precisely controlled using a depth monitor that operates based on optical interference. This step pretty much determines the profile of the poly–Si.
3. Then the bias is increased, and at the same time the gas ratio is adjusted to enhance the sidewall protection film. Next, 50–60 nm of poly–Si is etched with a high poly–Si/TiN selectivity of around 20. In this step, poly–Si stringers remain along the fins.
4. An overetch process, with an isotropic component added, is then used for etching off the poly–Si stringers with a higher poly–Si/TiN selectivity.
5. A highly anisotropic etching is used for etching the TiN partway, and then a process with some component of isotropy and a high-TiN/high-κ selectivity is used for etching the remainder of the TiN.
6. After the high-κ film is etched partway by dry etching, the remaining high-κ film is removed by wet etching. Wet etching offers a high selectivity to the underlying SiO_2. Figure 6.20 shows the etching result [15]. A vertical profile is achieved (Fig. 6.20a), and there are no etch residues along the fins (Fig. 6.20b).

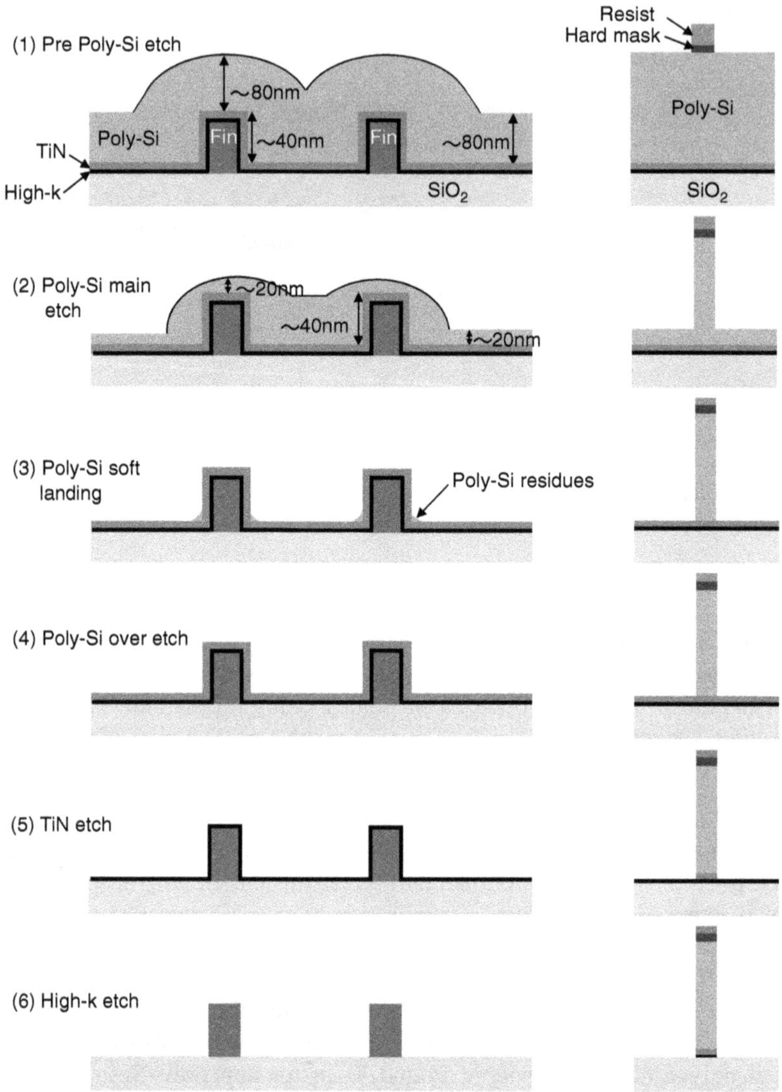

Fig. 6.19 Process flow for metal gate/high-κ etching for FinFETs [15]

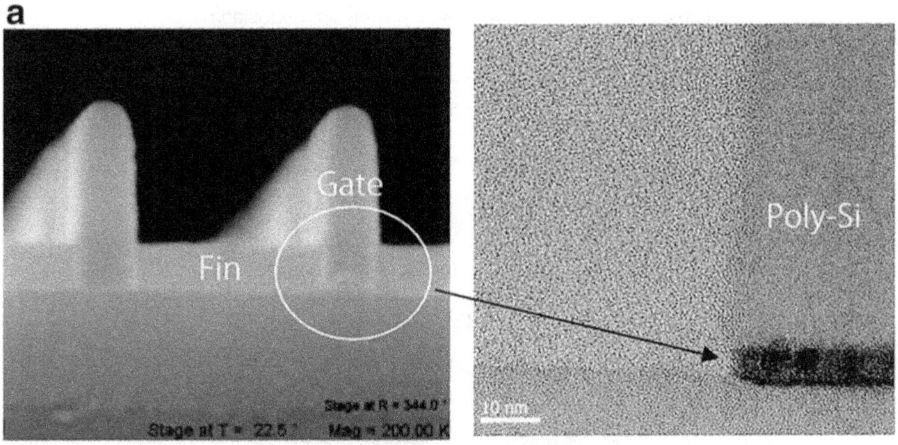

Scanning electron microscope cross-sectional profile of metal gate/high-κ

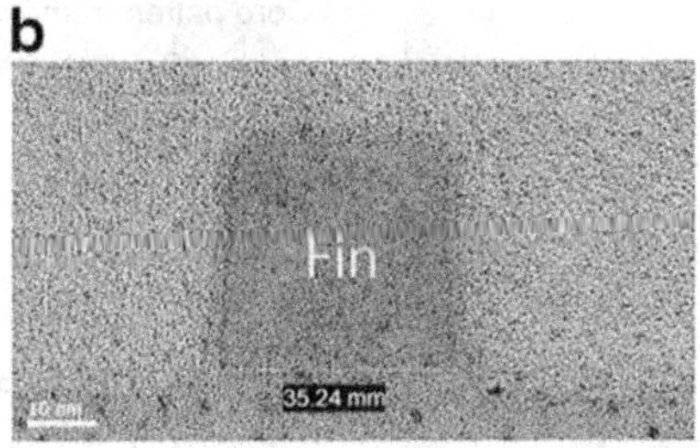

Scanning electron microscope cross-sectional profile of fin

Fig. 6.20 Etching profile of metal gate/high-κ for FinFET [15]

6.6 Double Patterning

The resolution in lithography technology is proportional to the wavelength of the light source. In other words, a light source with a shorter wavelength must be used for forming finer patterns. The commercially available lithography technology for forming the finest patterns today is the ArF immersion lithography technology. An ArF excimer laser (wavelength: 193 nm) is used as the light source, and the space between the lens and the wafer is filled with a liquid in order to increase the refractive index, so that even finer patterns can be formed. ArF immersion lithography technology is in actual use for production of 32-nm-node LSI. Even with this ArF immersion lithography technology, it will be difficult to form the patterns for the 22-nm node and beyond. Extreme ultraviolet (EUV) lithography technology is a promising candidate for the next-generation lithography technology. In EUV lithography technology, a soft X-ray of 13–14-nm wavelengths is used as a light source. However, the EUV lithography technology faces many issues, and its commercialization has been significantly delayed. As an alternative, double-patterning technology (DPT) is being considered. There are several schemes of double

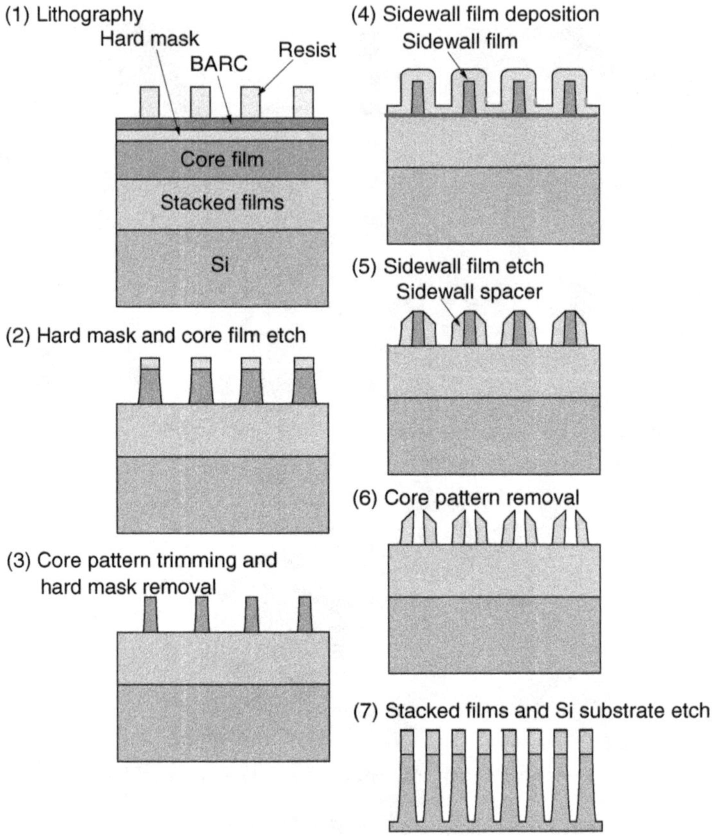

Fig. 6.21 Process flow for double-patterning technology [16]

patterning. The discussion here focuses on DPT using a sidewall transfer process, which has been commercialized for volume production of leading-edge flash memory devices.

Figure 6.21 shows an example of sidewall transfer DPT applied to Si trench etching (STI) [16]:

1. First, the stacked films, core film, and hard mask are deposited on the Si substrate, and resist patterns are then formed using ArF immersion lithography technology.
2. Next, the hard mask is etched, and this hard mask is then used as a mask for etching the core film.
3. The core patterns are trimmed to make them thinner.
4. A sidewall film is deposited over the core patterns.
5. The sidewall film is then etched anisotropically by dry etching, leaving the sidewall spacers.

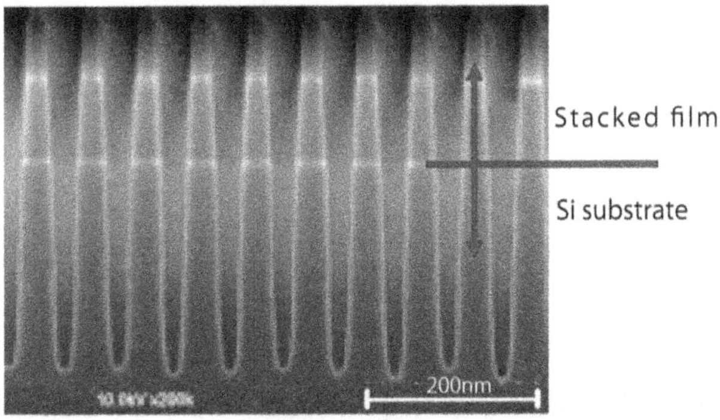

Fig. 6.22 Scanning electron microscope cross-sectional profile of 32-nm line and space pattern formed by double-patterning technology [16]

6. The core patterns are removed so that only the sidewall spacers remain. At this point, the sidewall spacer pitch is one half the resist pattern pitch in step (1). In other words, if the resist pattern pitch in step (1) is 96 nm (48-nm line/48-nm space), then the sidewall spacer pitch obtained in step (6) would be 48 nm (24-nm line/24-nm space).
7. The stacked film is etched using the sidewall spacers as a mask, and then the Si substrate is etched for completing the STI formation.
8. As thus shown, the sidewall transfer DPT makes it possible to create fine geometry patterns that are one half the pitch achieved at the lithography resolution limit.

The process flow in Fig. 6.21 shows that core film etching [step (2)], core pattern trimming [step (3)], and spacer etching [step (5)] have an impact on the mask CD and pitch. In this way, in the DPT, dry etching determines the accuracy and variability of mask CD. The materials used for the core film and sidewall film are selected to achieve a large selectivity to the spacer as the core patterns are removed, and for the right selectivity to the underlying films. Figure 6.22 shows a cross-sectional SEM photograph of the 32-nm line/32-nm space STI formed by DPT [16].

6.7 Etching Technology for Three-Dimensional Integrated Circuits

The commercialization of EUV lithography technology, considered to be promising as a next-generation lithography technology, is significantly delayed, and current lithography technology has come to limit the speed of device scaling. While DPT can be used as a temporary solution until EUV lithography technology is commercialized, there are several issues. First, the sidewall transfer DPT described in the preceding section only works for repeating patterns such as those found in the flash memory devices and cannot be used for random patterns in logic devices. Random patterns require double patterning with exposures in two steps. Second, double

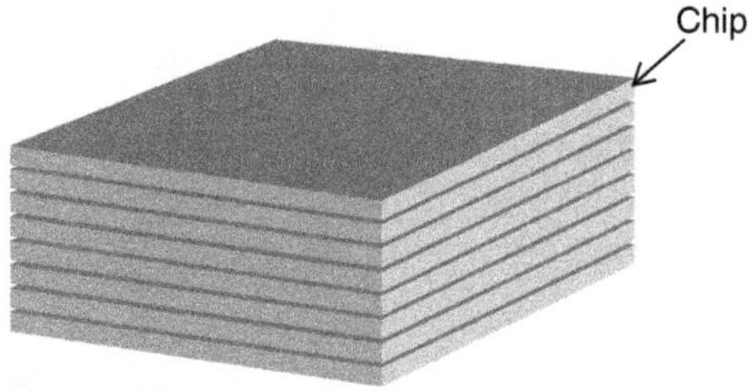

Fig. 6.23 Three-dimensional integrated circuit (3D IC) by chip stacking

patterning requires a large number of process steps and the manufacturing cost increases. Also, even with double patterning, the lithography technology will hit another limit in resolution with further device scaling. Even if EUV lithography technology were to be commercialized, the cost of equipment would be extremely high, and it may in turn increase the manufacturing cost. Device scaling itself may also be about to hit its limitation because of issues of electrical characteristics. For example, flash memory becomes susceptible to errors when the amount of stored charges decreases with scaling. In logic devices, the variability in transistor performances due to the transistor dimensional variability is no longer negligible.

A lot of research work is going into three-dimensional integrated circuits (3D IC) as a means for overcoming these issues [17, 18]. As shown in Fig. 6.23, a multitude of chips are stacked together to increase the device density with this technology. The density of memory devices such as DRAM and flash memory is increased by stacking a multitude of memory chips. While a system LSI integrates the heterogeneous devices like logic and memory onto a single chip in the conventional system on chip (SoC), with the 3D IC, the logic chip and memory chip are stacked together to create a system LSI. Figure 6.24 shows a cross-sectional structure of the 3D IC.

Through-Si vias (TSVs) are used for chip-to-chip connections in the 3D IC. The TSV is deep, of approximately 50–100 μm, and the etching technology used for forming the TSV is called deep-Si etching. An etching method called the Bosch process is generally used for TSV etching. The Bosch process consists of a repetition of polymer deposition and Si etching, taking place one after the other. It is a type of sidewall protection process. As mentioned in 2.3.3 in Chap. 2, polymer deposition and etching take place at the same time in a sidewall protection process. With the Bosch process, polymer deposition and Si etching take place alternately. The process sequence is shown in Fig. 6.25:

1. First, Si is etched to a certain depth with SF_6.
2. Then polymer deposition takes place with C_4F_8 to protect the Si surface.
3. Polymers at the bottom surface are removed by etching, and then Si etching continues. Polymers remain on the sidewalls and prevent the radical attacks. Then these steps are repeated to form the TSV.

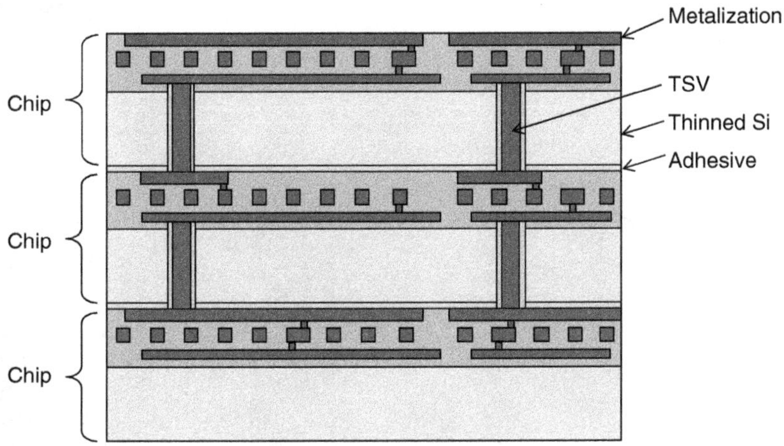

Fig. 6.24 Structure of 3D IC

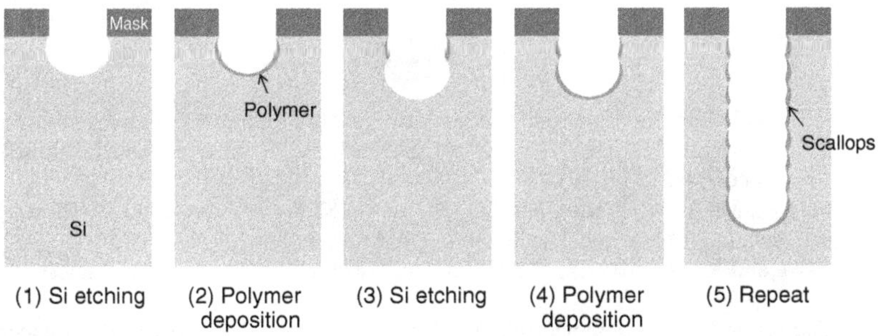

Fig. 6.25 Sequence of Bosch process

The time required for polymer deposition and etching is usually around 1 s. The Bosch process results in wave-shaped roughness, called scallops, on the sidewalls. It is necessary to minimize the scalloping as much as possible with TSV etching. Some of the effective means of achieving this are to shorten the polymer deposition and etching times and to reduce the amount of side etching during Si etching. Figure 6.26 shows a cross-sectional SEM photograph of the TSV formed by the Bosch process [19]. The diameter and the depth are 5 μm and 40 μm, respectively. There is almost no visible scalloping.

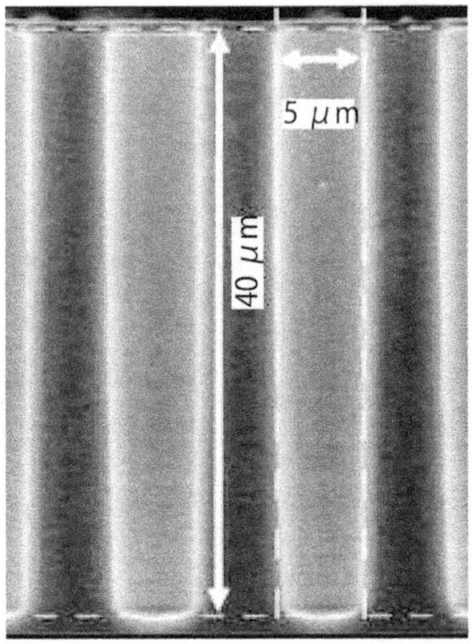

Fig. 6.26 Scanning electron microscope cross-sectional profile of TSV formed by Bosch process [19]

References

1. M. T. Bohr: Tech. Dig. Int. Electron Devices Meet., p.241 (1995).
2. S. Uno, T. Yunogami, K. Tago, A. Maekawa, S. Machida, T. Tokunaga and K. Nojiri: Proc. Symp. Dry Process, p.215 (1999).
3. S. Machida, A. Maekawa, T. Kumihashi, T. Furusawa, K. Tago, T. Ynogami, T. Tokunaga and K. Nojiri: Tech. Dig. Advanced Metallization Conf.: Asian Session, p.19 (1999).
4. T. Furusawa and Y. Homma: Ext. Abstr. Int. Conf. Solid State Devices and Materials, p.145 (1996).
5. H. W. Thompson, S. Vanhaelemeersch, K. Maex, A. V. Ammel, G. Beyer, B. Coenegrachts, I. Vervoort, J, Waeterloos, H. Struyf, R. Palmans and L. Forester: Proc. Int. Interconnect Technol. Conf., p.59 (1999).
6. M. Fukasawa, T. Hasegawa, S. Hirano and S. Kadomura: Proc. Symp. Dry Process, p.175 (1998).
7. M. Ikeda, H. Kudo, R. Shinohara, F. Shimpuku, M. Yamada and Y. Furumura: Proc. Int. Interconnect Technol. Conf., p.131 (1998).
8. K. Mistry, et al: Tech. Dig. Int. Electron Devices Meet., p.247 (2007).
9. K. Choi, et al: Tech. Dig. Symp. VLSI Technology, p.138 (2009).
10. G. Kamarthy, I. Orain, Y. Kimura, A. Kabansky, A. Ozzello and L. Braly: Proc. Symp. Dry Process, p.47 (2009).
11. K. Ono, K. Takahashi, K. Eriguchi: J. Plasma Fusion Res., 85, 185 (2009).
12. K. Nojiri, G. Kamarthy and A. Ozzello: SEMI Technology Symposium (2009).
13. H. Kawasaki, et al: Tech. Dig. Int. Electron Devices Meet., p.237 (2008).
14. H. Kawasaki: Oyo Buturi, Vol. 79, No.12, 1103 (2010).
15. G. Kamarthy, G. Lo, I. Orain, Y. Kimura, R. Deshpande, Y. Yamaguchi, C. Lee, and L. Braly: Proc. Symp. Dry Process, p.43 (2009).
16. K. Yahashi, M. Ishikawa, H. Oguma, M. Omura, S. Takahashi, M. Iwase, H Hayashi, I. Sakai, M. Hasegawa and T. Ohiwa: Proc. Symp. Dry Process, p.279 (2008).
17. R. Dejule: Semiconductor International, p.14, May (2009).
18. P. Marchal and M. V. Bavel: Semiconductor International, p.24, August (2009).
19. C. Rusu: 55th Int. Symp. America Vacuum Society, PS2-FrM5 (2008).

Chapter 7
Future Challenges and Outlook for Dry Etching Technology

7.1 Innovations in Dry Etching History

The author joined the semiconductor industry in 1975. Combined with his 3 earlier years spent on semiconductor research in college and in graduate school, the author's total involvement with semiconductors has spanned more than 40 years. During that time, semiconductor technology has made incredible advances in terms of device scaling, density increases, and larger wafer diameters. As mentioned in Chap. 1, the device minimum feature size has been shrunk by approximately 30 % every 3 years, and the LSI device density has been increasing pretty much according to Moore's law. In 1975, when the author first started in the semiconductor industry, a large number of discrete bipolar transistors were still in manufacturing, and 16K DRAM, based on the 5-μm process, was just about to begin. As of 2014, devices with 20-nm-level minimum feature sizes had gone into volume production. This means that minimum feature sizes were scaled to 1/250 in 39 years. At the same time, the number of transistors on each microprocessor chip grew by approximately 100,000-fold. Furthermore, the Si wafer diameter, which used to be 75 mm in 1975, had grown to 300 mm. It is a different world today.

Dry etching technology, along with lithography technology, has been a key technology driving device scaling. In 1975, when the author started his career, wet etching was the mainstream etching technology, with some amount of dry etching in the barrel-type equipment used for resist ashing, wafer backside film stripping, and etching of the insulating film at the bonding pads. At that time, however, Hosokawa et al. at ANELVA were already working on reactive-ion etching (RIE) research, which eventually led to the RIE dry etching equipment for volume production. Their RIE technology received the 25th Okouchi Memorial Prize [1]. As a result, dry etching technology took its first step for fine geometry processing. This was the first technological innovation in dry etching.

K. Nojiri, *Dry Etching Technology for Semiconductors*,
DOI 10.1007/978-3-319-10295-5_7

The next technological innovation came with the development and commercialization of the single-wafer dry etching equipment. As described in great detail in Chap. 4, single-wafer processing had to replace batch processing in order to keep up with the expanding wafer diameters, and this required a technology for forming a high-density plasma at low pressure. The electron cyclotron resonance (ECR) plasma etcher, developed by Suzuki et al. at Hitachi, realized this, and the author was also involved in that work. This technology was recognized with the 36th Okouchi Memorial Prize. In fact, it received the Okouchi Special Prize, considered the most prestigious among all of the Okouchi Memorial Prizes [2].

The plasma used for dry etching causes charging damage, and at one time there was a concern that this problem could preclude the use of plasma for further scaling. Nevertheless, a great deal of active research work, mainly in Japan, took place in the late 1980s through the first half of the 1990s on understanding the charging damage, and researchers eventually gained a complete understanding of the issue [3]. Author's group at Hitachi, where the author was located, and groups at Toshiba, Fujitsu, and Panasonic did extensive work on measuring and modeling the damage. It is not an overstatement to say that Japan led the world with this work, and a multitude of issues were resolved as a result. In the end, plasma technology continues to be in use today, ant it will continue to be used into the future. In that sense, this work was one of the key technological innovations in dry etching.

The history of semiconductors includes a large number of technological innovations, not just in fine geometry processing, and the industry has commercialized many technologies that used to seem out of reach. As a result, it has achieved the prosperity seen today. There will be many more new technologies coming their way to continue driving the advances in the semiconductor industry.

7.2 Future Challenges and Outlook

As mentioned in Chap. 6, the commercialization of extreme ultraviolet (EUV) lithography has been significantly delayed, and double patterning is used as a temporary solution. Furthermore, quadruple-patterning technology (QPT), with two repetitions of double patterning for cutting the pattern pitch to 1/4, is being studied. Nevertheless, the devices themselves are also about to hit their limitations in scaling because of their electrical performances, and LSI technology will surely migrate to 3D structures. In addition to the chip stacking technology using through-Si via (TSV), as described in Chap. 6, flash memory technology with 3D stacked memory cell arrays is also being proposed, creating new challenges for etching technology. For example, the bit-cost scalable (BiCS) flash memory technology developed by Toshiba [4] requires an etching technology capable of opening high-aspect-ratio holes in a single step through stacked layers consisting of poly–Si and the insulating film that are stacked one after the other.

New types of memories are explored as an alternative to DRAM and flash memory, which suffer from a reduced number of stored charges as a result of scaling.

Some of the new memory technologies being considered are phase change random access memory (PRAM), magnetoresistive random access memory (MRAM), and resistive random access memory (ReRAM). A new material such as GST, consisting of Ge, Sb, and Te, is used for PRAM. Magnetic films, such as NiFe and CoFeB, and tunnel barrier films, such as MgO, are used for MRAM. ReRAM consists of various types of metal oxide materials. Therefore, new etching technologies will need to be developed for these new materials. The materials used for MRAM are especially difficult to etch because it is difficult to obtain volatile etch byproducts. An innovation will be required not only with the etching process, but also with the hardware for preventing the nonvolatile byproducts from depositing onto the chamber walls.

Silicon etching technology will be required to meet two contradicting demands. First, a high etch rate will be required for etching deep holes as in TSV. On the other hand, the etching technology used for gate etching must offer a low etch rate because of the very thin gate film and the need to control the amount of Si recess (the amount of Si substrate that is etched away). Recent research has shown that the Si recess will lead to a decreased drain current in MOS transistors and shifts in the threshold voltage V_{th} [5]. It will be necessary to reduce the amount of Si recess to almost 0. Therefore, the gate etching technology of the future will require atomic-level control such as an atomic layer etching, in which one atomic layer is etched at a time. It will be impossible to meet these contradicting demands using a single etching method. Different types of etching equipment optimized for each type of process will be required.

As devices are scaled, they will, of course, require more stringent specifications on contamination and damage. The physical damage to the Si substrate will be an issue, and low-ion-energy etching will be required. One of the potential solutions is the atomic layer etching. The requirements on the metallic contamination on the Si substrate also become more stringent, and it will be necessary to use even higher-purity materials inside the etch chamber or apply a coating of high-purity film on the materials.

The wafer diameters will expand to 450 mm. The plasma source for the 450-mm process, of course, must meet a high uniformity requirement. It is thought that a simple scaling up of the conventional plasma sources will no longer work adequately. A new type of plasma source will need to be developed.

7.3 What Engineers Should Do

For some time, semiconductors have been called the "bread and butter of the economy." Semiconductors will become even more essential in the information society. The semiconductor industry will definitely continue to grow, and the future of semiconductors is bright. While device scaling is facing many different types of limitations, the 3D structures that will overcome these limitations are nearing commercialization. This author believes that there will always be new and innovative technologies that will break the barriers. The author experienced such a breakthrough firsthand through his research work on the ECR plasma technology and charging damage. The author has gained an extensive experience working in production fabs, and there were no problems that could not be resolved with focused efforts by the people working there.

The technological demands on the dry etching process are becoming more stringent, and process development is becoming more challenging. The conventional approach of optimizing the process conditions through repeated experiments and reliance on the engineers' experience and intuitions will no longer be enough. It is necessary to build a process using the insights into how the plasma behaves and the reaction mechanisms. Not everything is known yet about the reaction mechanisms that take place during dry etching. As mentioned in Chap. 2, there is a handful of basic data and concepts that may be applied, for example, for selecting the gas chemistries, based on the ion energy dependence of the chemical sputtering yield and atom-to-atom bond strength. What is necessary is always to go back to the basics. It is easy to think of the etch chamber as a black box, but engineers need to remember to think about the fundamental scientific principles and consider what is happening in the etch chamber.

Engineers engaged in etching technology development must also know about the peripheral technologies, or other process technologies. They also need to deepen their understanding of device structures and device characteristics. For example, knowledge in both plasma and devices is required in order to understand plasma damage and come up with effective solutions.

What is most important is the pioneering spirit that drives one to go after new challenges. Perseverance is also important. Then any problem can be resolved by bringing people together so that they can share their knowledge and wisdom. It takes time to develop new technologies and new materials and bring them to commercialization. Executives and managers need to understand that and support the efforts that young engineers make.

Plasma will continue to be in use for fine geometry processing. The author hopes that younger people will make their careers in this field.

References

1. T. Tanaka, K. Hanazawa, K. Ukai, N. Hosokawa: Report of 25th Okouchi Memorial Prize (1979).
2. K. Suzuki, Y. Kawasaki, Y. Kakehi, K. Nojiri, S. Shimizu: Report of 36th Okouchi Memorial Prize (1990).
3. K. Nojiri: Charging Damage in Semiconductor Process, p.51 (REALIZE INC, 1996).
4. H. Tanaka et al: Tech. Dig. Symp. VLSI Technology, p.14 (2007).
5. K. Eriguchi, A. Matsuda, Y. Nakakubo. M. Kamei, H. Ohta, K. Ono: IEEE Electron Device Letters, 30 (7), p.712 (2009).

GPSR Compliance
The European Union's (EU) General Product Safety Regulation (GPSR) is a set of rules that requires consumer products to be safe and our obligations to ensure this.

If you have any concerns about our products, you can contact us on

ProductSafety@springernature.com

In case Publisher is established outside the EU, the EU authorized representative is:

Springer Nature Customer Service Center GmbH
Europaplatz 3
69115 Heidelberg, Germany

www.ingramcontent.com/pod-product-compliance
Lightning Source LLC
Chambersburg PA
CBHW050806100925
32286CB00003BA/178
* 9 7 8 3 3 1 9 1 0 2 9 4 8 *